THERAPEUTIC TOUCH AS TRANSPERSONAL HEALING

THERAPEUTIC TOUCH AS TRANSPERSONAL HEALING

Dolores Krieger, Ph.D., R.N.

Lantern Books • New York
A Division of Booklight Inc.

2002
Lantern Books
One Union Square West, Suite 201
New York, NY 10003

Printed in the United States of America

Library of Congress Cataloging-in-Publication Data

Krieger, Dolores.
Therapeutic touch as transpersonal healing/Dolores Krieger.
p.cm.
ISBN: 1-59056-010-8 (alk. paper)
1. Touch—Therapeutic use. I. Title.
RZ999.K753 2002
615.8'51—dc21

2002003256

TABLE OF CONTENTS

ACKNOWLEDGMENTS

Writing this book has served to bring together many of my thoughts over the thirty years since my colleague Dora Kunz and I first brought Therapeutic Touch into being. The intervening years, during which she and I continued to develop Therapeutic Touch as a healing way, have been keenly challenging but thoroughly satisfying. This was in no small part due to the demanding but nevertheless thrilling exposure to the thoughts of Emily Sellon and F. L. Kunz, Dora's husband. Looking back over those thirty years, I no longer recognize the person I was three decades ago, a salutary effect for which I thank these three people most sincerely.

I also want to express my appreciation to Gene Gollogly of Lantern Books for his encouragement and enthusiasm for *Therapeutic Touch as Transpersonal Healing*, and to Sarah Gallogly for her insightful editing of the manuscript. I am thankful for the excellence of Loren Wheeler's photography, and grateful to Stacey, Janet, Jim, Caroline, and Denis for the use of the candid photos taken of them during the Therapeutic Touch process. I would also like to acknowledge Maria Arrington: friend, tireless typist, and discriminating critic. Finally, I would like to acknowledge the thoughtful feeding and watering by my sister, Oh Shinnah Fast Wolf, who took care that the animals—wild and domestic—and myself were regularly fed during the nine months of intensive but rewarding labor that birthed this book.

FOREWORD

JEANNE ACHTERBERG, PH.D.

Transpersonal healing has been used for at least as long as medical artifacts have been identified and dated, which is over 20,000 years. The cave paintings found in the south of France provide an example of visions of animal spirits used for healing purposes. But despite the ancient origins of transpersonal healing practices, the term "transpersonal healing" and the evolving healing system it represents are relatively recent developments.

About ten years ago, having observed many subtle changes over the years in the practice of mainstream (allopathic) medicine, I recognized the presence of an evolving foundation of research, a theoretical infrastructure, and methods of diagnosis and therapy that were generating a new system of healing. In order to breathe more life into these new healing tools, and to integrate them into standard care, a name was important. I coined the phrase "transpersonal medicine," believing it would serve as an appropriate descriptor of the territory, although other terms, such as transpersonal healing (as used in this book) may be far more fitting. (The word "medicine" creates a cascade of feelings and opinions—and many assume that doctors alone can practice it. In fact, "medicine," in tribal and modern language, refers more broadly to that which helps or heals, or the art and science of treating and preventing disease and relieving suffering.)

"Transpersonal," a term with variously claimed coinage, was popularized by Antony Sutich and Abraham Maslow to describe a fourth "force" in psychology. The first three forces, psychoanalysis, behaviorism, and humanistic psychology, are characterized by a singular intent to focus on the inner life of an individual and the external circumstances that provoke changes in behavior or consciousness. Transpersonal psychology looks beyond the development and expression of the self to the greater, more cosmic, global, and spiritual concerns of the human community. Use of the term has now moved beyond the sphere of psychology, as anthropologists, ecologists, and others employ the term "transpersonal" to describe a similar thrust in their own fields.

In the medical field, transpersonal concepts mark the pinnacle of the evolution of the modern health care system. Our starting point is allopathic (Western) medicine, the underlying assumption of which is that disease is an entity that usually has some external cause. Treatment is focused on removing the cause or alleviating symptoms via surgery, pharmaceuticals, and technology (diagnostic, monitoring, life-support, radiation, and replacement of body parts).

The relatively new system known as mind/body medicine assumes, in contrast, that behaviors have something to do with health. Behavior is variously defined, from observable activities such as eating, smoking, exercise, work, and so forth, to a broader concept that includes internal events such as attitudes, emotions, and responses to stressors. Behavioral modification, biofeedback, nutrition, exercise, counseling, imagery, relaxation or meditation, expressive therapies (such as music, art, and dance), a variety of stress management techniques, hypnosis, body work (such as massage), the martial arts, vocational counseling, and health education represent a short list of the tools and techniques used. Mind/body medicine is practiced by a horde of professionals, including psychologists, physicians, nurses, biofeedback therapists, counselors, and allied health specialists.

Transpersonal healing represents a further step, extending beyond the development and expression of the self to the human community, as well as to the Divine, the Creative Source, God, the All—or whatever language is used to describe spirituality. One of the basic assumptions of transpersonal healing is that "something"—usually something invisible—is transferred between and among individuals, as well as from a spiritual source and from living and nonliving objects that serve in a healing way. That something can be referred to as energy, prayer, dreams, visions, or distant intentionality. Many practitioners of transpersonal healing also believe in the healing power of community, of ceremony, and of place or "sacred space." Practitioners of transpersonal healing vary from professionals who perform Therapeutic Touch, as developed by Dolores Krieger, to native or indigenous healers, to members of church congregations, "energy healers," and more. Ritual or ceremony and symbols are critical, serving as the tangible, physical representation of what is believed to occur in the invisible realm. Many systems of healing, including traditional Chinese medicine, Ayurvedic medicine, homeopathy, and medicine practiced by indigenous people such as Native American, African, and Hawaiian, have transpersonal assumptions built into their precepts.

The common element in most transpersonal healing is that both patient and healer enter into a state of consciousness that is unlike the usual, wide-awake state associated with active thought. The state is often referred to as a trance, an altered or nonordinary state of consciousness, a meditative mind, or some special brainwave frequency, such as is associated with alpha or theta brain waves.

The patient's experience varies with the technique used. As in allopathic healing, patients may be passive while something is done to them. They may be prayed for, touched, or placed in the center of community ritual, such as in Navajo and Tibetan sandpainting healing ceremonies. Or the patient may take an active role by seeking healing from his or her own prayers or intentions, using age-old imagery techniques, affirmations, pilgrimages to sacred sites, or through dreams and visions.

The definition of what currently constitutes medical practice is more fickle than absolute and is clearly in flux. Psychologists are negotiating for prescription and hospital admission privileges, and acupuncturists and chiropractors have been given the status of primary care providers. There has been some slow progress in integrating the evidence for mind/body interactions into health care: diet, nutritional status, phytopharmacology, and vitamin/mineral supplements have been researched and relatively accepted in the prevention and treatment of cancer; and mind/body medicine has been especially well accepted in areas where regular medicine acknowledges limitations, such as cardiac rehabilitation and chronic pain.

In this book, Dolores Krieger provides a foundation for the next step: the frontier and future of health care. She has written a scholarly treatment of Therapeutic Touch and transpersonal healing, with theory, empirical studies, and clinical applications that will be useful to a vast array of professionals. Readers who have had some experience with Therapeutic Touch will certainly reap great benefit, but the book will be valuable to all those who desire support and direction for healing modalities that derive their benefit from invisible and unknown (as yet) channels of information transmittal. Like all of Krieger's work, the book is both intuitively inspired and clinically supported. It is sure to be regarded as one of her major literary contributions.

INTRODUCTION: WHAT YOU ARE ABOUT TO READ

The basic assumption of this book is that you, the reader, have had previous experience with Therapeutic Touch (TT) through practicing it or reading about it. Specifically, *Therapeutic Touch as Transpersonal Healing* has been designed for the TT therapist who is committed to helping or healing those in need and who has found entrée through this process to deeper dimensions of herself.* It is out of these profound levels of consciousness that Therapeutic Touch arises.

As the therapist incorporates the TT process into her daily life and becomes cognizant of her potential for expanded perceptual ability, she frequently wants to pass this opportunity on to others—either by teaching them about the TT process, by serving as an exemplary model of its practice, or by mentoring novices who share her mission. This book is not intended to be a formal pedagogic text but a basis for the understanding, appreciation, and sharing of the subtle but informed context of Therapeutic Touch from which the TT therapist draws her working principles.

To accomplish this, the book examines three aspects of this healing interaction. Part I accompanies the reader through an in-depth study of the conceptual frame of reference that underlies both the theoretical content and the practice of Therapeutic Touch. Part II is concerned with the TT process, its interpretation, and how the therapist's understanding of both informs her personalized TT techniques. Part III discusses various principles of teaching and how they pertain specifically to Therapeutic Touch. Insight questions are offered, along with suggestions for discussion of the broad implications of TT for both healer and healee.

In the Appendices, you'll find sources of information about the continued education of teachers of Therapeutic Touch, a collection of references, a concise listing of significant principles to keep in mind when conducting a Therapeutic Touch workshop, a self-evaluation test, and other material of interest to the TT teacher or the advanced TT therapist.

In contradistinction to my previous writings on Therapeutic Touch, this book has been written with purposeful formality and depth. I wished to highlight the underlying elegance and cohesiveness of the Therapeutic Touch process as clearly and explicitly as possible for the benefit of the clinicians, teachers, and researchers who will continue to develop Therapeutic Touch in the future. I have also tried to keep in sight the simplicity and availability of the TT process. I have enjoyed this challenge and hope that you, the reader, will enjoy the result.

*Throughout this book, the pronoun "she" is given preference in referring to the Therapeutic Touch therapist, since healing is a recognized function of the feminine principle.

PART I

THE CONCEPTUAL FRAMEWORK FOR THERAPEUTIC TOUCH

OVERVIEW

Therapeutic Touch (TT) looks exceedingly simple. Most often it is only through the mature practice of it that one realizes that TT is an absorbing dynamic process of unsounded depth. Moreover, both the TT therapist and the healee engaged in this multidimensional process are unequivocally open systems of vibrant, human energies, each set clothed in its own intentionalities, enactments, and visions.

To gain a valid perspective of the forces at work during the TT process, one has to step back from the vital engagement of human energy systems and seek out the underlying living frame of reference that makes it possible. One way to do this is to examine the set of concepts and the body of ideas that my colleague Dora Kunz and I called upon in conceiving TT as a contemporary interpretation of several ancient healing practices that have continued to be effective even in this time of high technologies and rapid cultural change.

The presentation of a conceptual framework for TT will serve many purposes for the teacher and for those who want to delve deeply into the healing process. It will furnish a context for TT by specifying the elements that are similar to or at variance with Western science, and by highlighting areas that will require further clarification at some future time, due to a current lack of research tools or understanding. It will also reinforce the future development of theory and indicate viable directions for new studies of the Therapeutic Touch process.

As background, a short essay, "What is Healing? A Brief History," will introduce the discussion of the conceptual framework of Therapeutic Touch. Subsequent chapters will include a review of

- The significant concepts we have used in the development of TT, and what they mean
- How these concepts are used in the practice of TT as a healing modality
- How the comprehension and appreciation of these concepts integrate within the TT therapist as they are put into practice
- The effect of these dynamic interplays on the healee
- Indicators of a conscious realization by the TT therapist of the farther reaches of the therapeutic processes in which she is engaged

This parsing out of the TT process will serve as an overview, helping to focus the scope of multifaceted actions in which the therapist is dynamically engaged during the healing act.

Since the best learning mode of the TT process is experiential, the definitions that will be used in developing the Therapeutic Touch conceptual frame of reference have been arranged roughly in the order in which the TT therapist experiences them while she is engaged in the TT process. To take advantage of this structure, it is suggested that the reader take a moment to center and then visualize her most recent experience of TT. If the reader can pace her visualizations to her reading of the various definitions, she will have at hand a ready example of how the concepts come alive during the TT session, and the following pages will become more meaningful.

CHAPTER I: WHAT IS HEALING? A BRIEF HISTORY

Healing as a Way of Power

Healing is a way of power, a door to the paranormal and, therefore, to the mysterious. It is an entry-point to the sacred, for healing has a vital concern with the sanctity of life, a consideration that can even show us how to appreciate the grace of death.

The earliest practices of healing others were those of the shaman—particularly the female shaman, for it is thought that the prototypic healers were women who sat around the home fires in prehistoric times and tended the ill and the wounded. The concept of the shaman as a feminine role is supported by the custom that it is the woman who assures the survival of the children and the children's children. However, the recognition that the ability to heal gave power to the healer led to the institution over time of complex rituals to safeguard healing's seemingly secretive nature. As history approached the edges of modern civilization, these rituals were eventually taken over by the dominant religion, and the vocation became more male-oriented.

The ways of healing reach back in time to the earliest historical records. Pictographs etched on the walls of the Cave of the Three Brothers in the Pyrenees depict healing by the laying-on of hands.

The Ebers Papyrus, dating back to 1552 BCE, mentions the use of this technique to treat injuries to the head. Cuneiform tablets uncovered from the sands of the Euphrates River tell of the healing ability of Trita, a physician of ancient Persia, and Indian texts hoary with age attest to the powers of the famed ancient physician Dhen Wantari.

Folktales about healers and feats of healing defy time in a continuity of traditional tellings and retellings down through the generations. In ancient Greece, Hippocrates (c. 460–357), the most famous of physicians in the Western world—although almost nothing is known about him beyond a brief reference by Plato—is purported to have used the laying-on of hands as a common mode of healing. Around 400 BCE, the laying-on of hands also was reported by Aristophanes, the famous Greek playwright. The ancient use of this technique, for the curing of snakebite in particular, was later confirmed in the writings of Pliny the Elder, a Roman statesman and historian who lived at the beginning of the common era.

The Aesculapian temples of healing in early Greece bore on their walls inscribed testimonies of the laying-on of hands as a treatment for blindness and infertility. These temples were also made famous by their use of incubation, a healing process by which the cause of a patient's illness is revealed to him or her during sleep. Accounts of this process often seem similar to what can happen to an ill person today, as the modern-day therapist helps the client in uncovering unconscious origins of his illness. The records indicate that the ill person was met at the door by an attendant, who washed him and clothed him in a clean, shift-like garment. The attendant would tell the patient something of this nature: "Tonight you will sleep with the gods, and in your sleep they will tell you the causes of your illness." The patient would then be given a meal laced with a sedative, which would cause him to fall into a deep sleep. When he awoke from the sleep, he would find a therapist sitting by the bed who would help him recall his dreams, and the therapy would begin.

Early Roman texts cite Galen of Greece, who attempted to systematize the whole of known medicine, for his ability to heal. Galen was physician to Marcus Aurelius, the Roman emperor (161–180 BCE) who was also well known as a healer. In fact, several warriors who became emperors were known for their healing abilities. Of particular distinction was the emperor Vespasian (69–79 CE). He was known particularly for his ability to intervene therapeutically in cases of neurological disorders, lameness, and blindness. The emperor Hadrian (117–138 CE) was said to be able to reduce "dropsy," a condition of abnormal accumulations of fluids in the body tissues and cavities that can accompany heart disease, cirrhosis of the liver, and some kidney diseases. Known today as edema, it is one of the symptoms that can be quickly and significantly reduced by treatment with Therapeutic Touch.

The high regard in which healers were held often spilled over into political affairs and other power relationships within the community. Consequently, healers, or persons who purported to be healers, became leaders in local and regional government. In Europe, both religious and political authorities resented the healers' rise to power. Early in the eleventh century, there was a political power play between church and state: both wanted the healer's power for their own—the church as it began to interpret the cause of illness as sin, and royalty so that they could use healing to emphasize the notion of divine right. As a result, a certain type of healing, usually modeled after the laying-on of hands, was specially assigned to the nobility and dubbed the Royal Touch. Particularly in England, France, Germany, and Norway, kings and other noblemen sought to assume this power by declaring themselves healers.

The Royal Touch

The notion of the Royal Touch was quickly accepted, and the early kings and queens of England and France became renowned for their ability to cure goiter and other throat ailments. The noblemen of the House of Hapsburg were thought to be able to cure stammering with a kiss! People who were healed by the Royal Touch were often given a token, such as a gold coin, which they proudly wore around their necks to indicate royal favor.

In England, the Royal Touch started with Edward the Confessor, an Anglo-Saxon who reigned from 1042 to 1066. It is recorded that Edward the Confessor was able to cure throat ailments and epilepsy, for which he was canonized by the Roman Catholic Church in 1161. It is an interesting side note that in *Macbeth* Shakespeare has Malcolm and Macduff take refuge in the court of Edward the Confessor, where Malcolm, to his amazement, witnesses a miracle. He reports:

> . . . strangely visited people,
> All sworn and ulcerous, pitiful to the eye,
> The mere despair of surgery, he cures,
> Hanging a golden stamp about their necks,
> Put on with holy prayers; and 'tis spoken,
> To the succeeding royalty he leaves
> The healing benediction.
>
> (Macbeth, IV, iii)

The tradition of the Royal Touch continued in England for about seven hundred years, until its popularity weakened, faltered, and finally petered out during the reign of Queen Anne (1702–1714), the last of the Stuart monarchs. Among those she touched was Samuel Johnson, made famous by his biographer

Boswell; nevertheless, Johnson is said to have "died from the clap" (gonorrhea).

In France, the tradition of the Royal Touch extended from the time of Philip I (1060–1108) to the reign of Louis XIV (1638–1715), who was said to have touched 1,600 people one Easter Sunday. Although much of the tradition of the Royal Touch was clearly sham, healing others could be a deeply moving experience, evoking attitudinal change in some royal healers. Notably, there was King Olaf of Norway (later the country's patron saint), who took healing seriously and was canonized in 1164, shortly after Edward the Confessor of England (1161).

Healing and the Church

The laying-on of hands is common worldwide as a method of healing that has been performed most frequently by clergy, and infrequently by laymen of the dominant religion. In Christianity, Matthew says that Jesus "went round the whole of Galilee teaching in their synagogues and proclaiming the good news of the Kingdom and curing all kinds of diseases and illness among the people." Jesus' disciples were given "authority to overcome the devils and cure diseas" (Luke 9:1–6). The New Testament contains forty-one citations of healing.

By the fifth century CE, St. Patrick was healing the blind by touch in Ireland, and by the tenth century, St. Bernard is known to have healed a wide variety of infirmities—blindness, lameness, muteness, and deafness—in the congregations of Cologne. During the next two centuries, healing continued within a religious context.

During the medieval period, "natural" healers—those who were able to heal without having been officially taught and accepted as healers—were often killed, or lived in the shadow of that threat, while priests and exorcists held the moral right to heal and practiced without interference. The state rarely intervened to protect "natural" healers; in fact, rulers who adhered to the notion of the Royal Touch often killed, tortured, and maimed lesser folk who were able to heal, imprisoning them on the charge of desire to usurp the throne. For the one who simply wanted to help or heal those in need, the exercise of compassion came at a very high price.

Perhaps another reason that those outside the clergy were denied the right to heal, particularly through the use of touch, was that the laying-on of hands has long served as a ceremonial tool. For instance, it is used in both Judaism and Christianity during the ordination of priests, rabbis, and ministers. It is also used to purify holy water or sacralize certain objects, and it is part of the ceremonies of baptism and confirmation, the blessing of congregations, and exorcism. As priests asserted themselves as healers, they came to be regarded as the only keepers of miracles. Nevertheless, in Europe several "natural" healers outside the clergy became well known.

Laypersons Who Could Heal

Valentine Greatrakes

In the West, the recognition of laypersons who could heal and could convince authority of the validity of their treatments has had scant notice until quite recently. Brian Inglis brought to my attention the work of Valentine Greatrakes, seventeenth-century Irish landowner and magistrate. Since he is one of the few healers in Western history who have been viewed in a positive light, it is useful to review his record.

After some years spent in the wars with Oliver Cromwell, Greatrakes returned to his home in Ireland and, for unrecorded reasons, became interested in the laying-on of hands. With practice, Greatrakes found that he could produce a therapeutic effect without actually touching the ailing body. (This fact surprised me, for I had thought that healing without direct body contact was something novel we had discovered and developed in Therapeutic Touch!) This was very important for Greatrakes' time, since it obviated the need to remove a patient's clothing for treatment.

This was the time of the feudal states, and Greatrakes spent much time healing the peasants who worked for him. Word of his success spread locally, then regionally. Among the records of his healings were instances of paralysis, deafness, and headaches. It is said that the swellings of tumors and the inflammation of arthritis also disappeared under his touch.

In 1666 (about five hundred years after Edward the Confessor initiated the Royal Touch in England), Greatrakes had a highly successful healing tour of London. Colorful reports of his arrival at the docks of London[1] make it easy to imagine the festive mood and the excitement with which he was greeted, the many-hued flags that overarched the streets, the gay playing of instruments, and the exuberant welcome that greeted him.

Since Greatrakes was popular and a member of high society, he could demonstrate his abilities without being accused of dealing with the devil. Other circumstances helped protect him as well: he had published a description of his methods, including detailed case histories; he freely admitted that his methods were not always successful and publicly stated that the healing force could only come from God; and, being a man of some wealth, he refused payment for his treatment.

While these factors kept him safe from the accusations of witchcraft that were common at the time, it did not earn him the respect of physicians. On the London tour, they refused to pay attention to him, in spite of his many requests to be tested by them, and they silently ignored his findings. Greatly downcast by this refusal to take his healing work seriously, Greatrakes quietly returned to his estate, leaving the medical profession—and, more importantly, the people—all the poorer. Upon his retirement to his estate, Greatrakes was soon forgotten—but not by me. In fact, I look upon Greatrakes as the patron saint of Therapeutic Touch.

Don Pedro Jamillo

Quickly passing over the next few centuries—during which, nevertheless, the lives of several healers acted as models for their neighbors or otherwise influenced the period—I would call your attention to one who was a legend in his own time, the Mexican charismatic healer Don Pedro Jamillo, who lived in Southwestern Texas during the nineteenth century.

His healings and demeanor were such that he became all but a myth among the Border people, and his fame spread beyond those borders. After his death he was apotheosized and became an internationally acknowledged folk saint in whose name many healings are claimed even today.

Olga Worrall

One healer whose attitude exemplifies that of most healers through the 1950s and '60s was Olga Worrall, perhaps one of the most intensely studied healers of our time, who died a few years ago. Olga's approach demonstrates how, even by the middle of the twentieth century, healing was regarded almost exclusively within a religious framework.

Olga said that she was born with healing ability and went into the healing ministry early in life. She gave herself to innumerable researches out of her religious convictions about healing. One of her most remarkable feats—under carefully controlled conditions—was altering from a distance the hydrogen bonding of water, making it "heavy water," by using her healing abilities on an experimental sample.

She explained her ability thus: "We work with God's laws of perfect health . . . in energizing the atmosphere around a person." She continued, "I just let God take over . . . I become a clear channel, God-centered . . . I am in a constant state of prayer . . . I tune in and then become a clear channel and let the energy flow."[2]

Therapeutic Touch: A Contemporary Pioneer in the Teaching of Healing

In 1972, my colleague Dora Kunz and I founded Therapeutic Touch. Its reliability and validity were tested with extensive research,[3] and, as its credibility was established, it was adopted in the health sciences as an extension of their professional practices. By the end of the twentieth century, one hundred medical centers and health agencies were including the practice of Therapeutic Touch in their in-hospital and outpatient services.[4] The master's level course we developed in 1975 for New York University marks the first time in the recorded history of healing that healing per se has been consistently and formally taught within an accredited college or university curriculum. At the university's suggestion, I named it "Frontiers in Nursing." It continues to be taught at NYU at this writing, and its curriculum has become a model for graduate classes in the colleges and universities of many countries throughout the world.

As of this writing, Therapeutic Touch has been taught in more than eighty foreign countries. A listing of information on seventy of the schools where Therapeutic Touch is taught in North America, and a listing of about seventy-five medical centers where Therapeutic Touch is presently practiced in North America, can be found in the appendix of a previously written book.[5]

The Need to Heal

This brief history of healing demonstrates that the urge to become a healer may arise from many disparate motivations. If invoked as an act of will to power, healing may play surrogate for a need to exert influence, authority, or control over dependent persons. The healing act may be a demonstration of faith, and in some belief systems it may be used as a ceremonial tool during certain rites in the quest to be at one with God. But it seems that sympathy, empathy, or compassion most often underlies the motivation for this most humane of all human acts.

These last-mentioned traits are not considered to be necessary survival skills in the long evolution of man. Why, then, do they persist? My growing hunch is that there are significant but unrealized gaps in what we know about the human condition. One of these gaps contains a latent drive, a need to help or heal that remains close to the surface of our personalities but has been strongly repressed in Western society at various times by both church and state.

Abraham Maslow, one of the seminal thinkers at the forefront of the humanistic transpersonal psychology movement of the 1960s, believed that our species was still evolving toward a fully human stature. He conceived a theory of needs—what we require for survival—and metaneeds—"higher" needs that must be filled if we are to progress toward that ideal of a "fuller" humanity.[6] (These metaneeds rest on a person's "B-values" or Being-values—values that support our decisions about life events.) Failure to fulfill these needs, he said, leads to "metapathologies," or "sickness of soul."

"A sickness of soul"—even though historically the soul has defied attempts to define it, the phrase evokes a state we can easily imagine. And indeed, a "sickness of soul" does envelop those who ignore the call to sympathize or empathize, or to be compassionate, and who later realize they have lost an opportunity to aid fellow human beings. In fact, even a memory of such a denial has the power to plunge one once again into " a sickness of soul" in anguish for the missed opportunity.

In reading Maslow, I realized that in his theory of metaneeds lay one rationale for the question: "Why do people want to help or heal others in need?" The healer's journey actually allows her to realize her metaneeds. Surveys have indicated that most people who have become healers have followed a persistent quest for meaning in life. This resolute seeking leads to an increased sensitivity to self and others that can evoke spontaneous (that is, creative) inner responses to crises—whether our own or someone else's. The press of psychic energy involved at this critical point can cause an upwelling of desire to give or receive help.

If one's aspiration to help or heal goes beyond the sphere of personal needs, there comes a moment of ego transcendence in which the mere desire to help becomes a driving need, a compelling desire to protect others from the suffering we have known. In Maslow's view, this would constitute a critical shift of identity from the nuclear or primary family to the family of Man, a significant change in worldview that can carry a person beyond the basic needs of mere survival to the recognition and appreciation of metaneeds. Compassion and concern for others are B-values that are foundational to the metaneeds necessary for the fulfillment of our potential as human beings. In other words, compassion is a prime, high-level strategy for wellness, for its realization brings us further along the path to becoming more fully human.

Within this context, an answer to the question: "Why do people want to help or heal others in need?" becomes possible. The need to help or heal is not to be dismissed lightly; this metaneed arises out of the unconscious depths of human thought and feeling. People want to heal those in need because that is the way human beings should act toward one another to fulfill their own destiny. To heal others is to make oneself whole, to have high-level wellness of soul, and to share that sense of wholeness with others who are in dire need. Healing is a high calling that speaks in the language the inner self understands.

CHAPTER II: CONCEPTS AND CONSTRUCTS ESSENTIAL TO THE THERAPEUTIC TOUCH PROCESS

Balance

The various subtle elements of the healee's vital-energy field are constantly in motion, in a sustained state of rhythmic pulsation that is normally in continuous interchange with the environment. This state of continuous interaction is considered to be healthy when it is in balance, and unhealthy or indicative of illness when unbalanced.

Basic Assumption

Accepted without proof; taken for granted.

Bilateral Symmetry

Matching formations, networks, or conditions that are opposite to each other over a boundary. In the TT assessment, the spinal column is used as the central longitudinal axis to determine equal distribution, balance, or harmony within the healee's vital-energy field.

Centering

The point of entry for the TT process, a state of consciousness that is maintained throughout the TT session. In a sense, the act of sustained centering in TT is the ground against which the TT techniques act as figures.

Centering is an act of self-searching, a state in which the TT therapist quiets spontaneous surges of consciousness and undertakes an intentional exploration of the deeper levels of her self and their relationship to the universe. In the process of becoming aware of that relationship, the therapist has the opportunity to recognize slumbering potential in herself that can be consciously awakened and actualized.

Chakras
Translated from Sanskrit as centers of consciousness whose structures are non-physical. They are natural components of human-energy field dynamics that can transform universal energies into human energies. These subtle structures are reflected at the various levels of consciousness of the individual person, but they have their source as universal fields of mind.

Charisma
The ability to inspire followers with devotion and enthusiasm.

Compassion
To help or to be merciful to one who is suffering; a deeply felt drive to help or to heal someone who is in need.

Cue
A subtle feeling of difference in the healee's vital-energy field that is perceived by the hand chakras of the TT therapist during the TT assessment and the rebalancing of the healee's vital-energy field. Once the region is rebalanced, the cues, having nothing distorted to respond to, no longer relay information in a manner that can be perceived by the therapist.

Directing Vital Energy
A technique of the TT process in which vital energies are projected with specific intentionality from and through the TT therapist to areas of deficit, depletion, or debilitation in the healee's vital-energy field.

Dying Process
A form of change of physical, emotional, and mental states of being; a transformation in consciousness.

Field
A field is thought of as a "condition" in space that can exert a force even in the absence of any physical, material medium, but which is not itself directly affected by the presence of any such medium.

Within the domain or province of the field itself, a sphere of influence is exerted which is an uninterrupted continuum of effective force. This field force is conceived to have an inherent binding power that is specific to the character of the field under consideration.

Field Counterpart
A non-physical complement of a physical organ, tissue, etc.

Grounding
Psychological term meaning to help stabilize and balance a healee's surging emotional energies; i.e., those occurring in hysteria or other unstable behavior.

Healee
Person in a state of imbalance who is therapeutically treated by a healer.

Healing

In general, healing involves the conscious, full engagement of the healer's own energies, plus other universal energies to which she has access, in the compassionate interest of helping those who are in need. Therefore, healing can be said to be a humanization of energy in the interest of helping or healing another person.

Human Energies

Force, vigor, or capacity for activities, behaviors, or functions innate in the human being. The range of human energies includes physical actions, emotions, feelings, reasoning, abstract thought, aesthetics, aspirations, and expressions of spirituality.

Human-Energy Field

This term is a metaphor. Energy fields accept as test objects things or living beings that can relate to them and can demonstrate physically or indicate the inherent characteristics of that field. For instance, iron filings placed in a localized magnetic field will indicate by the position they assume the north (positive) and south (negative) polar regions of that field. Moreover, the individual iron filings can be tested to determine the strength of the field forces holding them in their grip along the herzian lines that demarcate the invisible force exerted by that field. Filings of lead or slivers of wood do not have these abilities and therefore would not make good test objects to elicit the otherwise invisible distinctive characteristics of this localized magnetic field.

Similarly, one can assert that the vital-energetic, psychodynamic, and conceptual elements that constitute human qualities, such as emotions, abstract thoughts, aspirations, etc., have the ability to respond to the human-energy fields and, in so doing, to reflect the indwelling characteristics of those fields as they act out in the daily life of the person, the individual test object.

(See also *Levels of Consciousness*, below.)

Identification
Attributing characteristics of a model to oneself.

Imbalance
Indicates that the healee's vital-energy field is in a state of incoherence or deficit, or is impeded in some way, causing the vital-energy flow to become dysfunctional. When a person's vital-energy flow is in a state of imbalance, he is ill, traumatized, or overly fatigued.

Inner Self
Refers to the timeless link between the personality and the spiritual; the Guide or Teacher. Characterized by quietude, peacefulness, clarity of mind, certainty, heightened confidence, joyfulness, and an abiding sense of inner strength.

Intentionality
Deliberation, purposefulness. During the TT process, intentionality implies that the healer is driven by a particular goal in addition to the will-to-power, desire, or hope that inspires her actions.

Levels of Consciousness
Each fully formed human being has direct access to the universal (species-specific) human-energy field and therefore can actualize potentials that correlate with a certain gradation or range of sensitivities to various states of consciousness. These states describe and distinguish the universal human-energy field.

Proceeding up a scale of acuity from a base of fundamental physical awareness, human emotions constitute a second level, or subfield, of consciousness. We will refer to this level as the psychodynamic field of the individual. We act out the characteristics of this field through our personal emotions and our rationality.

Beyond that sphere of consciousness is that person's conceptual field. It is from the matrix of the conceptual field that we are enabled to conceive or to create, to aspire, or to act with intentionality. At the farther reaches of consciousness appear to be those personal qualities that make up the deep center of one's individuality, the realm of the inner self (see above). The spiritual level of consciousness is the individual's link with the universal base of consciousness itself.

Listen

To give attention with the ear; to be aware by listening alertly. In Therapeutic Touch, "listening" refers to the state of intent and inner awareness to which the TT therapist attends during the assessment of the healee's vital-energy field.

Metaphor

The application of a name or a descriptive term or phrase to an object or action to which it is imaginatively but not literally applicable.

Modulating Vital Energy

Redistribution of the healee's vital-energy flow to establish synchronous rhythms or harmonic patterning. The modulating of vital energies in Therapeutic Touch includes methods of sedating, vitalizing, quickening, or strengthening the multilevel streams of vital energies.

Open System

An organized complex of things, organs, or events that allow access and communication.

Overdose (OD)

Going beyond the point of balance of the healee's vital-energy field while transmitting healing energies to a healee.

Symptoms in the healee may include increased restlessness or signs of heightened irritability and intensified anxiety (e.g., hostility, anger, pain, or fear).

Patterns

As energies flow at the molecular level where quanta become matter, they have a natural tendency to gather into groups, strata, or zones so that a regular or logical form, order, or arrangement occurs in the media through which they stream. It is in the brain that patterns and levels of energy become meaningful and can be conceived as multi-dimensional models for more abstract thought or states of being.

Placebo Effect

Physiological or psychological responses to the ingestion or injection into the body of inert substances of no known therapeutic value; psychodynamic factors, such as suggestion, persuasion, coercion, and expectation, may be involved.

Prana

The Sanskrit term designating the potent energy flow (see *Vital-Energy Field*, below) that is used most often during the TT process and is the basis of all life fields. Prana is one of the most powerful energies in the vital-energy field; it vivifies, vitalizes, and animates all living beings. It can be conceived of as a nonphysical, energy-rich, unifying environment that interpenetrates every living cell and actuates its functions.

Presence

An individual's force of personality.

Principles of Order

The most fundamental concept upon which Therapeutic Touch is based is that principles of order underlie the functioning of the universe. These universal principles prescribe or define the organization and behavior of all consciously directed actions and improvised or spontaneous happenings that occur in the known universe. Principles of order imply a predictability of events in which one may have a high level of confidence.

Rebalance
Refers to the rebalancing of vital energies in the healee's vital-energy field. During the TT process, the act of rebalancing is cue-specific; that is, those regions of imbalance or dysfunction in the healee's vital-energy field that were elicited as cues during the TT assessment are the basis for the rebalancing phase of the TT process, specifying the TT techniques that will be used and the therapeutic goals for this phase of the TT session.

Relaxation Response
The term was popularized by Dr. Herbert Benson's studies of Transcendental Meditation. It refers to a relaxed, meditative state of consciousness.

Repattern
To return the streaming of vital energies to a previous configuration of vital-energy flow.

Replicability
A statistical term indicating a highly consistent act that occurs repeatedly and cannot be realistically attributed to mere chance.

Rhythmicity
A measured flow of successive, pulsed motion.

Simultaneity

Occurring or operating at the same time, neither occurrence necessarily being either the cause or the effect of the event; that is, the importance of the event is relative to each individual's cognizance of it.

Telereceptor

A telereceptor senses information (such as light and sound) over distance without connecting neurons. For example, the senses of taste and smell act over distance through their responses to chemical molecules that carry stimuli through fluid and air. Sight occurs via photons of light, hearing occurs through acoustical pressures that carry over distance to impinge on the acoustical nerves, and even touch can occur without direct contact, as in Therapeutic Touch. In this healing way, the telereceptor for the sense of touch senses over distance via cues picked up by the TT therapist's hand chakras as they move through the healee's vital-energy field.

Therapeutic Touch

A contemporary interpretation of several ancient healing practices that have endured through time. Ancient healing practices that are incorporated into the TT process include the laying-on of hands, visualization, touch with and without contact, centering the consciousness, the therapist's knowledgeable use of certain of her chakras, and the intentional use of breath for therapeutic purposes.

Touch

To come into or be in physical contact with [another thing] at one or more points; a cutaneous sense; to palpate or manipulate. However, as used in Therapeutic Touch, touch (via the hand chakras) can be nonphysical.

Transcendence
The act of going beyond the range or the grasp of human experience, belief, or reason.

Transformation
The act of making a thorough or dramatic change in the form, outward appearance, or character of an individual; a sudden, dramatic change.

Transpersonal
Describes the experience of a state of being beyond the usual range of the personality; it is characterized by a sense of being an integral part of a whole.

TT Assessment
An evaluation of the healee's vital-energy dynamics based upon the TT therapist's direct experience of that individual's vital-energy field as perceived through the sensitivity of the therapist's hand chakras.

TT Reassessment
A rapid assessment of the healee's vital-energy field during or at the end of a TT session. Its purpose is to pick up clues about how to continue, or, if there no longer are cues, to recognize that it is time to conclude the TT session. In this way, the TT reassessment prevents the possibility of the healee overdosing on unused vital-energy build-up.

The Unconscious
The part of the mind that is not accessible to conscious thought but affects emotions, behaviors, and thinking.

Universal Healing Field

A multidimensional domain whose milieu is conducive to the reestablishment of a state of wholeness to injured, debilitated, or ill functions or faculties of living beings. It is considered universal because all living beings within this sphere of influence have a natural potential to heal themselves and others in need.

Unruffling Vital Energy

A type of modulating vital energies during the rebalancing of the human-energy field that gets at the structural quality ("texture") of the vital-energy field by using the hand chakras and physical movements to translate cues toward the periphery of the healee's vital-energy field, where they can dissipate.

Visualization

The pertinent dictionary definition is "to make visible, especially in one's mind (as a thing not visible to the eye)."

Vital-Energy Field

Vital energies are restless, shifting, rhythmic patterns in constant flow. These configurations of subtle energies are the basis for the physical functions, emotional and behavioral patterns, drives, thoughts, and intuitions that are uniquely encoded in each individual.

The vital-energy field is the personal multidimensional space that surrounds and quickens each individual, energizing and reinvigorating him or her throughout life. It is characterized by the dynamic flow of vital energies that derive from sources that are little understood in the West but will be introduced in later chapters.

CHAPTER III: CLUSTERS OF CONCEPTS AND CONSTRUCTS RELEVANT TO THERAPEUTIC TOUCH

Concepts are formulated to express the essential quality or nature of groups or classes of objects or events. They take into consideration all the aspects of the object or event. Constructs are words or terms defined, often by consensus, in a particular way, and apply only under specific circumstances.

The Case for the Healing Moment: Order

The concept of order—what the contemporary writer Samuel Clarke called "the eternal fitness of things" and the ancients called "The Great Chain of Being"— has always intrigued humans. The earliest known human thoughts were cloaked in awe at the evidence of the stars' consistency, their predictable patterns giving testimony to an invisible ordering that governs the functioning of the universe. The ordering principles that seem to underlie the tiniest as well as the grandest acts of Nature work with a fine, explicit detail that is stunning to witness. Newton, after conceiving his brilliant laws of motion and gravitation, looked back and asked himself, "Whence is it that Nature doth nothing in vain; and whence arises all that order which we see in the world?"[1]

This concept of a fundamental order lies at the heart of the TT process, for healing is essentially concerned with correcting a state of imbalance through the intentional reorganizing of dynamic functions, feelings, ideas, and motivation.

This wonder at the evidence of Nature's inner workings stirred in our ancestors fundamental questions and searchings for an understanding of how this universe functioned. In trying to conceive of the strategies used by the universe to maintain the integral interrelatedness evident in its functionings, the Holy Bible, for example, relates the idea of a ninefold celestial hierarchy standing between God and man. Although Therapeutic Touch does not have a particular religious context (at this time, Therapeutic Touch has been taught in more than eighty countries, and so it is quite possible that TT therapists belong to every major religion in the world) it is important for TT teachers and committed therapists to appreciate beliefs about healing common among the many religions of the world. There are too many unknowns about the inner workings of the healing process to permit undue skepticism of others' beliefs.

The Categories of Aristotle

The predictability of the disciplined functions of the universe and the sequenced events obviously guided by natural law led to the realization, for which Aristotle is given credit, that things and events fall into distinct categories and classes. Specific ordering principles were seen to be embedded in this system, reinforcing the seeming infallibility of natural law.

Respect for these natural laws impressed the religious and the scientific alike. Nevertheless, in the early years of the history of Christianity the limited scope of science was acknowledged, so that St. Augustine's observation that "There is no such thing as a miracle which violates natural law. There are only occurrences which violate our limited knowledge of natural law" was held to be the case until present times, particularly by those focused on a religious context. As our understanding of the universe has matured, we have learned an even deeper appreciation for the regularity of natural law. Albert Einstein speaks strongly of "the harmony of natural law, which reveals an intelligence of such superiority that, compared with it, all the systematic thinking and acting of human beings is an utterly insignificant reflection." Nobel Prize–winning physicist Werner Heisenberg also saw that order within a holistic context: "From the cosmic distances to which man can penetrate by means of modern technology, we see perhaps more clearly than from Earth itself the unitary laws whereby all life on our planet is ordered."[2]

Contemporary Conceptions of Order

The person who has had the greatest impact on the modern notion of order in the universe has been David Bohm, one of the world's foremost theoretical physicists. Bohm, who did much work on the fundamentals of quantum theory and relativity and the philosophic meanings they hold for contemporary conceptions about the universe, was deeply interested in the nature of human consciousness. One of his most stunning conclusions as he put his life's work into perspective was his proposition that there is a hidden order (suggested by the undeniable wave nature of matter itself) at work beneath the apparent chaos and lack of continuance of individual particles of matter at the quantal level. It was Bohm's genius to realize that there was an invisible realm, which he called the implicate order, out of which the visible, physical substance of our space-time universe flows. He called this flow of the physical universe the explicate order; the folding and unfolding of matter that was characteristic of its actions he called holomovement.

Bohm's unique hypothesis was that it was from within the deeper folds of the implicate order that both matter and consciousness arise and become integrated as a unified field. One aspect of Bohm's thought that has immense significance for the comprehension of the healing act is that the source of immeasurable unbound energy is principally in the implicate order, which gives rise to physical, psychological, cognitive, and transpersonal human experience. The implications of this aspect of Bohm's theory will be developed further when we discuss the concept of centers of consciousness, or chakras, below.

Ordering Principles of Healing

Those who have had wide experience in helping or healing people who are ill are often impressed by what we might call ordering principles that inform the healing act. These principles, which assist in the reweaving and repatterning functions of healing, seem to be integrally rooted in the foundations of the healing process itself. Examples abound; we need only observe the healing of a wound—a small but deep cut on the hand—to note with wonder that it often heals without a scar or any other indication that anything once marred the intact tissue. Even at this gross level of direct observation, this would seem to be a valid indication of an intelligent organizing factor at work.

If we were to examine the healing process at a microscopic level, however, our appreciation would increase considerably. We would realize that healing involves several different levels of organization, each synchronizing its intrinsic functionings with incredible attention to detail. In order to heal severed or injured tissues, clouds of molecules of biochemicals specific to each of seven distinctly different cellular layers must be routed to correct sites on their hematologically active atomic space lattices as they mend and become functional again.

As we study the healing process more deeply, it becomes increasingly clear that certain less obvious but equally significant factors are dynamically engaged. Timing is one such factor. In the example of the healing wound, several biochemical substances must be brought into action at the molecular level on a schedule that is exquisitely synchronized. If the schedule goes awry, the tissue may become scarred or fail to heal. Time itself is known to be a significant and decisive factor in the restoration of form and function to damaged tissue, e.g., medical professionals know that with conventional treatment it takes six weeks for a fractured bone to heal, three days for skin adhesion to take place during wound healing, etc.

And yet, after thirty years of continual research and theoretical inquiry into the healing process, I am still awed by repeated demonstrations that time can seem to move differently during the healing act. I have seen organic change occur inexplicably but significantly faster than expected—on a few occasions noticeably occurring even under my hands as I was engaged in Therapeutic Touch—in cases concerned with wound healing, fractures, and the closing of operative sites. Also, there have been many instances in which patients appeared to experience unexpectedly rapid organic change in cases of cerebral vascular accidents, liver dysfunction, severe circulatory disturbances, urinary disorders, and peptic ulcers.

Unusually rapid healings are not infrequent in reports of TT therapists. However, what instigates the apparent acceleration of time in these cases evades attempts at logical comprehension. (My hunch, based on theoretical assumptions, is that there are critical enzymes highly sensitive to the healing process that act to speed up physiological processes—but definitive research has yet to be brought to bear on the question.)

Order and Meaning

While we may observe and marvel at the healing process, it does raise a provocative and profound query: Who—or perhaps what—is the master choreographer of these finely timed processes that are so intimately bound to the healing act? Whatever the "final cause," there is no doubt that the imminent availability of ordering principles and fundamental laws that we have yet to understand are our most unfailing allies in the healing act.

As physical expressions of reality, ordering principles appear to be at the core of the nature of the world itself. Such order seems to imply that there is purpose behind the organization of the functions of the universe. According to accepted logical thought, where there is purposeful organization, there is meaning. This suggests that as a function of the universe, life is meaningful, and that if it is meaningful, it can be understood. And, for the purposes of this book, if life's maintenance and regeneration can be understood, they can be taught and, most importantly, learned by those who would help or heal persons in need.

CHAPTER IV: THE UNIVERSAL HEALING FIELD

Field as Metaphor: An Analogue for the Healing Interaction

It was only in the last quarter of the twentieth century, after four hundred years of self-limiting mechanistic and reductionistic thinking, that Western culture became sensitive to a new vision of reality. In place of the cogwheel-within-cogwheel, push-pull, specific-cause-specific-effect idea of how the universe manages its day-to-day affairs, the emphasis of this fresh insight is on vital interconnectedness and relationship, dynamic patternings, and continual change and transformation.

Renowned twentieth-century philosopher of science Henry M. Margenau, in reviewing the history of relativity and quantum mechanics in the 1950s, noted that physical existence, as an indication of objective reality, is no longer limited "to what the eye can see." He cited James Clark Maxwell's concept of fields in physics, in particular his brilliant equations on relations between electric fields, magnetic fields, electric charges, and magnetic pole strengths, which opened scientific understandings of physical reality in unforeseen ways.

Physical Fields

With the extended development of theories of relativity and quantum mechanics during the twentieth century, the concept of physical (e.g., measurable) fields became more clearly delineated as it was more precisely defined. In the wake of these explanations, several conventions developed around its expanded interpretation.

A primary convention asserted that for appropriate use of the term "field," the context and the essential nature of that field must be rigorously defined. In the area of healing, there continues to be a multitude of unknowns, so a stringent definition of a "healing field" is difficult to structure. Because of this, one must accept that often the phrase "healing field" is used only as a metaphor.

The notion of field signifies a quality or a condition of a continuum, a specified domain that exercises a particular sphere of influence or control. The domain of a field may be invisible, but its effects are nevertheless perceptible, because they can be quantifiably tested. Although this sphere of influence is neither matter nor energy, it can exert perceivable influences or control over both. The gravitational field that attracts all material objects toward the center of Earth, for example, exerts an unseen but measurable force that affects matter and energy.

In theory, universal fields extend into space indefinitely. But, because it is non-material, a field does not occupy space, as do most objects we associate with physical reality. Consequently, innumerable fields can coexist concurrently in a specific space. The effects or characteristics of fields are perceived in objects that do occupy space—as in the destructive force of gravitation when an object falls to the ground and breaks. An object that is sensitive to specific field effects can be either qualified or quantified by appropriate and reliable measurement tools, and is therefore known as a test object for that field. One characteristic common to all fields is the ability to specifically organize or systematically arrange its test objects in the four or more dimensions of the space-time continuum, even at the molecular level.

The Basic Picture of Physical Reality

As an understanding of the concept of fields gained validity, it became the underpinning of a comprehension of both the macro-universe of the Einsteinian cosmos and the micro-world of quantum theory. By mid-century, Einstein, founder of the theory of relativity, could say, "We may regard matter as being constituted by regions of space in which the field is extremely intense. . . . There is no place in this new kind of physics for both the field and matter, for the field is the only reality."[1]

Over the years, the validity of the concept of field was severely tested and found to be an appropriate metaphor. By the last quarter of the twentieth century, Heinz Pagels, known for his creative work in quantum theory, could write the basic picture of physical reality within the relativistic quantum field theory frame of reference:

1. The essential material reality is a set of fields.
2. The fields obey the principles of special relativity and quantum theory.
3. The intensity of a field at a point gives the probability for finding its associated quanta—the fundamental particles that are observed by experimentalists.
4. The fields interact and imply interaction of their associated quanta. These interactions are mediated by quanta themselves, and
5. There isn't anything else.[2]

F. L. Kunz, in a succinct, integrated review of the major aspects of field theory, clarifies the non-physical nature of fields:

> Force field potentials are not seen. Nevertheless, they are known to exist; and what is more, they are understood, and their characteristics have been described in precise detail, although no one can directly experience them in the ordinary way through the senses . . . [the nature of field theory makes clear that] it is the non-material universe that is our true environment.[3]

Behavioral Fields

As the Post-Industrial Era blossomed into the Age of Technology, it was evident that the theoretical framework of physics had won the high regard of the rest of the sciences. First chemistry, then medicine and biology, closely followed by psychology and the social sciences, adopted concepts that originally had been conceived in physics and used them as metaphors in the development of theory in their own areas.

One such concept was field theory, and the universal healing fields of Therapeutic Touch can be understood within the context of that theory. In fact, with the accelerated rate of public acceptance of the ideas and practices of alternate healing modalities, and the increased appreciation for the notion of subtle energies involved in the healing interaction, the current use of the term "field" has taken on generally accepted experiential connotations as well as quantitative significance.

The Universal Healing Field in Therapeutic Touch

How the concept of the universal healing field developed vis-à-vis Therapeutic Touch is analogous to the concept's incorporation in the other psychosocial disciplines mentioned above. Therapeutic Touch is transpersonal in its healing effect and transcultural in its origins. Therefore, it deals with experiences that both transcend the personality and place the therapist in a state of interiority that can be deeply appreciated and understood by healers of many different lifeways. However, these states of consciousness are difficult to describe objectively in the manner prescribed by the physical sciences.

As used in Therapeutic Touch, the concept of the universal healing field involves a tacit understanding that there is a non-physical condition in space where space-time functions are so strongly and coherently integrated that all living beings who come within its sphere of influence are naturally impelled toward a healthy, unified sense of being. This so closely resembles the TT therapist's actual experience that each therapist soon adds personal conviction to her conceptual understanding of this field in which illness, injury, and dysfunction can become sound and healthy again.

This healing field, or sphere of healing influence, is universal because all living beings have the innate capacity to heal themselves. They also, we contend, have the natural potential to help or heal others in need. In this sense, the universal healing field is the source of all healing.

Within this context, the TT therapist is the test object for the universal healing field and the compassionate instrument of its innate healing force, which is translated to those with whom she engages in Therapeutic Touch. Within this impartial frame of reference, there is a twofold recognition that although the ultimate outcome of a TT session is not in our hands, nevertheless, when one steps forward to help or to heal someone in need, it is analogous to stepping through an open window in time to create a relationship in which one's inner self can actively participate. Therefore, the dictum for the TT therapist is that one should act upon a compassionate impulse if there is even a small chance of making that relationship therapeutic for someone in need.

Consequently, identifying with the universal healing field brings with it a particular state of consciousness, and the nature of that experience is oriented within a specific helping-healing reference frame.

This all-important state of consciousness will be further explored in subsequent descriptions of the Therapeutic Touch process.

CHAPTER V: THE HUMAN ENERGY FIELD: CONSCIOUSNESS AS A FUNDAMENTAL FORCE

We Don't Stop at Our Skin

The greatest gift of this "Newer Age" that was birthed in the 1960s is that it has given us permission to know ourselves. There is a significant quickening today of our collective realization of who we are as human beings, and occasionally we have had hints of the potential that is given us to actualize. It is an End Point, as Carl Jung termed it—not only the end of a century, or even of a millennium, but the end of an Age, the end of a mindset. It is important, therefore, that we learn to think in new ways; it is time for a radical shift in perception as we move into a new era, a new time of accelerating change and an opportunity to achieve an increasing recognition of the multiple realities in which we have our Becoming.

William Irwin Thompson reflected this insight: "It is time to use new senses, new perceptions . . . It is time to move into a new space." Christopher Fry said it a bit differently: "Will you wake, for pity's sake! . . . Now is the time to wake up. . . . We must act as if what we do makes a difference!"

The strength of our culture is that we maintain the freedom to explore our own potentialities. Now it is as though we are at the threshold of a new perception, a new cosmology of the universe "out there," and we thrill to indications of its intimate relationship to the subjective universe within. It is similar to the lunar cycles that are reflected in an equally mysterious way, not only in the tides of the seas, but also in the hormonal tides within our bodies, which carry in their cyclic ebb and flow the growth and development of the individual. On many fronts, the time is now opportune to permit ourselves to open to new, different insights and realizations of self.

One of the awarenesses that have been distinctive of this coming-of-Age has been a growing cognizance that "we don't stop at our skin." Western culture is not among the ninety-seven cultures that traditionally have acknowledged this as fact. In the West, Therapeutic Touch is a recognized pioneer in bringing this percept to professional and popular attention. While it was known previously that dynamic energetic interplays of physical forces reach beyond the body's boundaries, we in the West—particularly those of us who deal with people, their well-being, and their ills—in our fascination with the sophistication that this time of high tech lends us, had forgotten what the play of physics and chemistry throughout the body's functionings implies in reference to forces, energy nets, and the power of polar attraction and repulsion. We have only recently awakened to the fact that both non-living and living systems respond to energy. Energy, in fact, is the mother lode of our ability to be in the world. But, possibly because human energies are not readily viable, our culture has given them—and the effect they have on our humanness—scant attention.

Basic Physical Field Forces

Modern science recognizes that there are four basic forms of energy, sometimes referred to as field forces. These four field forces are assumed to account for all physical interaction. In our material universe, the force of the gravitational field is primary; it is the cohesive glue that allows matter even at the molecular level to stick together and become objects. The basic units of the gravitational field are called gravitons. It is the exchange of gravitons between all physical objects that dynamically affects, molds, and incessantly remodels the structure of the universe. In a more localized manner, the gravitational field also affects the human physical body, determining the stresses on its structure and significantly affecting its posture, locomotion, and all bodily functions, including energy flow.

The second basic form of physical energy is electromagnetic field force, which exchanges photons. Photons do not have mass and seem to be at the threshold where quantum mechanics operates, dipping down into the deeper dimensions of the implicate order. Light, which is the natural agent of electromagnetic radiation, is considered to be neither particle or wave, but is involved with both as it seems to "push up" and materialize in the world of the explicate order. At the molecular level, the interactions of the electromagnetic field (EMF) figure prominently in the physical and chemical properties of liquids, gases, and solids. At a grosser level, the EMF affects the fluid and electrolytic systems and the human nervous system.

There are two other recognized physical forces, the "strong" and the "weak" forces. The strong force, which is concerned with the interchange of gluons during nuclear reactions, is not now known to be significantly involved with human functions. Weak forces are demonstrable at sites of high radioactivity. Studies have suggested that these weak forces are present in instances of psychosomatic disorders, cancer, symptoms of high stress, endocrinal dysfunctions, and in some biochemical complications, such as diabetes. These weak forces also affect the body's fine structures at the molecular level, particularly where they are demonstrable at sites of high radioactivity.

In reference to healing in human beings, studies of weak forces in the last quarter-century have focused on the physiochemical linkages between mood, neuropeptides, and endorphins, which mediate pain and resist depression, and on relationships between endocrinal activity and the immune system. These multifactorial studies promise a significant breakthrough in our understanding of how the individual operates as a total integrated being, at least at the measurable physiochemical level.

The Life Fields

The development of an understanding of life fields is one way in which the concept of field is used in the biosciences. That life fields, or L-fields, have a directive and organizing function concerned with the physical structure of organisms was suggested in the 1930s at Yale University by Harold Saxton Burr and his colleague, Leonard Ravitz.[1]

Using microvoltmeters to test the life field of plants, Burr and Ravitz also demonstrated that a plant's—or another living organism's—state of health could be established in considerable advance of that organism's faintest physical signs. They also demonstrated that L-fields are true fields and not only skin resistance, for instance, by showing that when electrodes were placed at a distance from the surface of the body and not in contact with the skin, it was still possible to measure the voltage gradients in the "empty" areas beyond the subjects' skin; that is, L-fields shared the true field characteristic of action-at-a-distance.

Repeated experiments insured both the validity and the reliability of the notion that L-fields are crucial to the building, maintenance, and repair of all living forms. Gradually the work on plants was verified on human beings. Continued study by Ravitz demonstrated that the subject's state of mind significantly affects the voltage gradients of the L-field. This work was followed up by Edward Russell, who dubbed this indication of the power of thought the T-field, or thought field. He suggests that the T-field derives from what we know as the mind.

Human Subtle Energies

Beginning from these findings, the twentieth century has seen the development of a concept of subtle energies that are considered to be human. Among these are energies that connect us to the capacity for abstract thought, the ability to visualize and conceptualize, the aptitude to aspire to a goal, and the potential to engage the transpersonal. Such talents are considered to be evidence of distinctly human subtle energies and to derive from a human-energy field.

While all of the field forces affect human beings in some way—for each individual is an open system and therefore is accessible to all energy systems in the universe, and new forms of energy seem to be constantly on the threshold of discovery—uniquely human energies, such as those concerned with the deeper emotional and mental states, have been less intently studied in their own right. At present we understand how to measure states of consciousness validly only in reference to the four basic field forces. However, as previously noted, modern science now is willing to theorize about properties that can never be empirically determined (i.e., detected by the five major physical senses). Therefore, much work is currently underway in the study of subtle human energies, with researchers carefully fitting together pieces of information much as one puts together a jigsaw puzzle.

Interestingly, scientists have turned for direction to cultures that have developed methods of inquiry that are different from those of the West. For instance, in the West the bases of scientific inquiry are logic and objectivity, whereas inquiry in other cultures can be essentially subjective and may seem to be illogical, at least as far as accepted basic assumptions are concerned. However, it is on the basis of these latter, deeply experiential, personal knowledges that often have come to us from developing countries that much of our current understanding of human vital energies has arisen.

As noted above, it is estimated that there are at least ninety-seven cultures in the world at this time that recognize that each individual is surrounded by a human-energy field within whose domain surge patterns of vital energies in constant movement.[2] These patterns of vital energies flow rhythmically in specific configurations, and their supple structures are strongly correlated with the individual's functions and capabilities.

However, the psyche is very little understood in the West, and it wasn't until the latter part of the twentieth century that, having observed Nature's unaccountable ability to bring apparent order out of seeming chaos in either organic or nonorganic matter by repatterning events in a rational manner, the hard sciences in the West were forced to accept as a hypothesis a concept of mind as a significant variable. And so consciousness became a third factor, along with matter and energy, that had to be taken into account in considering the psychophysical world.

The subtle energy flows that underlie human function are called *rlun* by the Kalahari Bushmen of Africa, *sila* by the Eskimos of the Arctic, and *atua* by the Maories of New Zealand and the Aboriginals of Australia—a people whose ancient instruments and cave paintings give evidence of a wise and creative culture existing at least ten thousand years before the Egyptian dynasties.

It is often difficult for Westerners to accept that other cultures had sophisticated knowledge before the era of Christianity; nevertheless, it is historical fact that, in reference to healing, the histories of both India and China each go back approximately five thousand years in their studies of the essence of life. Both of these ancient cultures hold that the ultimate cause of illness is depletion, congestion, or imbalance of the constantly moving, rhythmic vital-energy flows (called *prana* in the Indian literature and *ch'i* in the Chinese). Of the two, the logic and assumptions of the Indian thought are more accessible to the Western mind, and so it has been the Indian frame of reference about subtle energies that has been used as model during the development of Therapeutic Touch.

As time progressed, our own experiences with TT reaffirmed for my colleague Dora Kunz and myself the validity of the Indian model, and deepened our appreciation of its depth. And in the quickening that has marked the last quarter of the twentieth century, there has been a decided movement within the life sciences to look to India and its ancient knowledge of the dynamics of subtle energies, in the hope that this traditional wisdom regarding universal laws will help with still unanswered questions about the several physical and emotional illnesses, mental aberrations, and spiritual emergencies that have not responded to Western medical practice.

CHAPTER VI: THE NATURE OF CONSCIOUSNESS

Toward a Realization of Human Consciousness

Therapeutic Touch has been called a "walking meditation"; that is, a conscious continuance of a state of meditation while the body is in motion. This is because the most potent and distinguishing feature of Therapeutic Touch is the centering performed by the TT therapist at the beginning of the session (and sustained for its entirety). Meditators of many traditions practice centering as a prelude to meditation, and as the practice of TT becomes a frequent and important part of the therapist's lifeway, she finds in the TT process itself the occasion and the permission to become more deeply involved in the centering act and the frequently concomitant meditative practice.

As the TT process continues and unexpected avenues of awareness reveal themselves with a clarity not previously experienced, an appreciation of the seemingly limitless aspects of one's consciousness heightens comparably. This stunning realization of unimagined potentials of expanded consciousness urges the therapist toward deeper studies of consciousness itself. Those who are involved with Therapeutic Touch realize over time that as their understanding of consciousness deepens, they can achieve more profound insights about the interactive aspects of the healing act. Below, I'll describe the role of consciousness both in human history and in Therapeutic Touch.

The Development of Consciousness

Consciousness has played a fundamental role in the history of human evolution, and it continues to do so as people mindfully strive to actualize their natural potential and fully realize themselves as human beings.

One can imagine that back in our primeval beginnings, human instincts were aroused to critical levels of awareness as our early ancestors were subjected to repeated physical impacts from the environment. Eventually, a more efficient nervous system evolved, enabling our awakening consciousness to be more sensitive in responding to the sensations we received. As these primitive people became increasingly aware and learned to separate awareness from immediate reflex response, they created in that pause a moment for reflection—a first glimmering of consciousness—on the best or most efficient response.

Over time the human body acquired a sensitivity to environmental stimuli that was based on the heightened response of several neurological systems. A fundamental activating vigilance was maintained by the reticular activating system in the more primitive base of the brain, and by the limbic system, whose functions are so submerged from our present consciousness that we usually are not aware of them until we are seized by deep emotion. In the labyrinth of the inner ear, a vestibular system developed to distinguish up from down in the space occupied by the body. The autonomic nervous system controlled or influenced involuntary functions, and, at the back of the brain, the layered cells of the cerebellum developed twelve different levels of information to coordinate and regulate muscular activity. Working together in exquisite efficiency, they constituted a body intelligence that assured the otherwise puny human body's survival among the fittest. Eventually, a consciousness that could be directed to probe countless questions about the universe became the crowning achievement and unique characteristic of human beings.

Aristotle and Descartes

It took a mind of remarkable depth to give logical structure to man's search for meaning. In the Western world, the pinnacle of this early quest can be found in the work of Aristotle, Plato's pupil, teacher to Alexander the Great. Aristotle separated all that was known about the universe into two distinct categories, the physical and the nonphysical. He further categorized them in a detailed summary that was held in highest esteem by Western civilization for the next two thousand years, until the present time.

In 1637, this classification was further clarified and more rigidly systematized into a mind/body duality by René Descartes, the French philosopher and mathematician, in his influential book *Discourse on Method.* The essential bent of his analysis of reality as man knows it was summed up in his famous dictum: *Cogito, ergo sum* ("I think, therefore I exist"), which reflected his concept of mind as conscious experience. An important implication of this statement was that personal consciousness is the only sure knowledge that the individual person has. This separation of reality into its communicable and objective components on the one hand, and its subjective and personal aspects on the other, continues to color the modern philosophical perspective.

The Evolutionary Potential of Communications Theory

In sharp contrast to a history of philosophic interpretation that reaches back for its raison d'être to the exhaustive classification systems of Aristotle and the dogmatic dualism of Descartes are the remarkably fast communication systems that have characterized the twentieth century, making prodigious amounts of information available on demand to people of almost any geographical location, racial derivation, and social class. With the new technology that permits equally instantaneous comparison and integration of these materials, we now have access to an awesome body of knowledge that incalculably exceeds the best education and understanding of previous generations.

One result of this has been the broadening of our consciousness of life and living as multifaceted phenomena. Our increasing sensitivity to the extensive number of interrelationships that engage us in daily affairs is inevitably expanding contemporary consciousness, giving it a richness of meaning and comprehension beyond anything previously imagined.

However, while the consciousness of our times is exceedingly broad, it is not necessarily of commensurate depth. For the past thirty years, the general level of understanding of the nature of consciousness has lain at the intersection between insights from modern physics and a spectrum of traditional teachings about meditation. More recently, however, the importance of unquantifiable qualities has been accepted, and it has been acknowledged that modern physics per se is not the appropriate avenue of enlightenment about human consciousness. At the time of this writing, rather than focusing on the study of consciousness as behavior only, significant consideration is being given to information theory and systems analysis of consciousness as a flow of experience (see The Study of Consciousness, below).

The Phenomenological Movement

Early in the twentieth century it was the phenomenologists, under the renowned German mathematician and philosopher Edmund Husserl, who captured the spotlight on the study of consciousness. The central concern of the phenomenological movement is the immediate conscious experience itself, rather than theoretical assumptions and explanations about the nature of existence. Phenomenology was the basis for gestalt psychology and was related to existential psychology, and the ties between these related disciplines strengthened and focused the drive to understand human consciousness as subjective fact.

In the meantime, many other avenues of inquiry provided intriguing findings about human consciousness. Some of the most fascinating were the studies done by the neurosurgeon Wilder Penfield, well known for his daring surgical work on the temporal lobes of people who suffered from epileptic seizures.[1] Penfield found that upon stimulation of certain areas of the surgical patient's brain, the patient remembered with amazing accuracy incidents that had happened in the past. The patients reported that they felt as if they were at the scene of the occurrence. Wilder later came to the realization that mind is not to be reduced to brain mechanisms, but is a significant factor in its own right.

Teilhard de Chardin's Omega Point

Another unexpected and significant finding that colors our current understanding of human consciousness was conceived by a French Jesuit paleontologist, Pierre Teilhard de Chardin. He perceived that it is our consciousness above all that imparts to each of us a vivid feeling of our livingness—a sense that we are ourselves and not other—the assertion that gives us a sense of identity. His studies, which reached back into the earliest times of living organisms, brought him to the conclusion that the promotion and development of consciousness was the primary purpose of evolution. This means that in actualizing our potential consciousness, we become more fully human.

Teilhard de Chardin introduced the concept of a noosphere, which he defined as the thinking layer of Earth—much as the Earth's biosphere is the layer of its atmosphere that contains the conditions for living. The noosphere, he theorized, was created by human intelligence, and has created conditions that now permit the extension of social and scientific capabilities in a manner entirely different from that of any previous evolutionary stage. Now, instead of the stimuli for man's evolution coming from his environment and other events that are outside his being, man's evolution will derive from his own consciousness, from "the within of things," as Teilhard de Chardin termed it. In effect, man's further evolution will be of heightened but complex unity, a state Teilhard de Chardin has termed the Omega Point.[2]

Tart's Taxonomy

Still another researcher whose studies have shaped our conception of consciousness is the American psychologist Charles T. Tart. In the 1960s, Tart pointed out that the human mind, or consciousness, has multiple significant facets, and that this high diversity is well known in other cultures—Sanskrit, for example, has more than twenty terms that are translated into English as various aspects of mind or consciousness. "Because in the West we do not have the vocabulary to specify the different shades of meaning"[3] that are embedded in these words, we continue to use the Sanskrit terms. Many of them, such as chakra, prana, and karma, have come into daily use in the discussion of realms of consciousness about which Western civilization remains ignorant. Where appropriate, the connotations of some of these terms will be discussed below.

Tart's research indicated that a wide variety of specific states of human consciousness exists. Out of these studies, Tart developed a taxonomy, or classification, of specific states of consciousness, each differentiated from the others. It includes the following independent states:

sensation	cognition
perception	intuition
emotion	self-awareness
affect	unition

Unition, a newly coined word, is defined as an experience of oneness or unification with everything.

Tart's taxonomy can be usefully applied to Therapeutic Touch:

- Sensation—verbalized in metaphors, e.g., heat, cold, electric shocks, tingling, during TT assessment
- Perception—includes intuition, mindfulness, cognition of subtle energies, etc.
- Emotion—becomes highly discernible, particularly when the TT therapist is assessing the areas over the heart chakra or the solar plexus chakra
- Affect—awareness increases as the TT therapist identifies with the healee's feelings
- Cognition—the TT therapist has the ability to translate nonphysical surges and patterns in the vital-energy flow of the healee into specific states of consciousness
- Intuition—increases in the TT therapist with heightened sensitivity to the inner self
- Self-awareness—the TT therapist is consciously aware of the TT transaction on both subtle and physical levels of consciousness
- Unition—a sense of timelessness and unity that can be experienced during the deep centering of the healer's consciousness during the TT engagement

The Study of Consciousness

According to psychology historian Eugene Taylor, the term "consciousness" is currently being used in three distinct ways. In research and philosophy, it is used as a theoretical construct referring to the system by which an individual becomes aware. Within a biological frame and in clinical practice, consciousness refers to reflective awareness, that is, one's awareness of being aware. Consciousness is also used as a general term encompassing all forms of awareness and intention.

Taylor explains that consciousness is measured in three different ways: phenomenologically, psychologically, and empirically. The real issue is that each of these ways of studying consciousness has a specific and separate method and context.

One way of studying consciousness is through direct experience, in which introspection, meditation, or free association is used. Another method is an observation technique in which the researcher uses her own consciousness to identify with the subject in order to understand the experience of that person or any other being. An example of this would be a study of cognition in children or, more recently, in animals. A third method of understanding consciousness involves making measurements of physical variables using input from any of the five major sense organs. An extension of this is used in Therapeutic Touch when the hand chakras act as telereceptors, probing the vital-energy field of the healee from a distance in their search for cues that indicate a state of imbalance in the vital-energy field.

There also are three ways of understanding consciousness, Taylor continues. There is the recognition of the primacy of consciousness in that consciousness exists as an experience "that transcends the split between the knower and the known by relating them." Secondly, consciousness can be considered as a field. In this context, consciousness is experienced "as a whole that unifies a multitude of perceptions, emotions and thoughts." It is then recognized that awareness occurs not from being aware of one thing at a time, "but rather as a unified field, in which certain aspects are attended to more than others." Finally, consciousness can be considered a stream. This concept is little changed from that put forward by William James, who introduced the term "stream of consciousness" to emphasize that consciousness is a flow rather than fragments of thought. In this context consciousness refers to the ongoing, ever-changing nature of consciousness itself as a unitary experience, rather than as a series of disconnected and unrelated events. The stream of thought includes the interaction of fragile, transitory images, a flow of sense impressions about the immediate environment, memories and reveries about former times, and anticipations and expectancies of the future.

The Power of Mind

As studies in the life sciences sought to understand deeper levels of brain function, here and there flashes of insight led to the recognition of the powers that the brain gains from association with nonphysical mind. Through the deepening study of psychosomatic medicine, particularly in reference to the interaction of the nervous system and consciousness, the life sciences were forced to recognize the unique human behaviors, such as the questioning mind, that were closely related to human consciousness.

The fundamental basis of psychosomatic medicine was Walter B. Cannon's model of the interaction of psyche and soma (essentially the emotions and the body) in what he called the "fight or flight mechanism."[4] In this model, the sympathetic nervous system mechanisms are stimulated by a fearsome experience and the body reacts to that fear with an instinctual behavior that forces the person either to flee the circumstances or to stand and contend with the alarming situation.

Hans Selye, who built on Cannon's work, attempted to bring together the biological and technological sciences with his far-reaching concept of overwhelming stress that inflicted damage on living tissues.[5] Selye conceived that stresses that continue over time initiate the fight-or-flight response noted above. This response then stimulates the hypothalamus, causing a chain reaction within the neuroendocrine system to muster energy for prompt action to meet the intimidating situation. As a consequence of the high degree of integration of the higher cortical and the lower subcortical areas of the brain required by this process, we are vulnerable to a wide spectrum of illnesses if the body, the mind, or the bridge between them is overstrained frequently.

Setting the Stage for Conceptual Revolution

The interest with which these findings were greeted triggered a profusion of rigorous studies of the mind-body relationship, which served to clarify differences and similarities between the physiological bases of stress, health, ecological cyclic behavioral influences, etc., and the neurophysical basis of states of consciousness. The end result has been a radical conceptual revolution, a recognition and acceptance that consciousness is at the foundation of all we experience, and as such is a fundamental feature of the universe alongside matter and energy.

What is not known to date, and remains a significant challenge for our time, is how matter, energy, and consciousness interact. One of the fascinating questions we face is whether consciousness, like matter and energy, is incontrovertible; that is, are matter, energy, and consciousness correlates, and, if they are, is there a simple equation for consciousness and its relationships, as there is in matter and energy: $e=mc^2$? One might even ask whether consciousness is a function of light, designated by "c" in the equation cited above, as the sages of another time hinted.

A few decades ago, such thoughts could not have been considered seriously. The radical change in mindset in recent years, and society's subsequent permission to release the consciousness of the people in new directions, has been the most significant happening of the twentieth century. Technology was a potent source of this transformation: the landing on the moon was a touchstone for dramatic and radical change in human regard for the potential of mind power. Close on its heels followed the introduction of the personal computer, which personally objectified for the user many of the operational functions of the mind and made the Internet widely available.

Perhaps technology's most stunning effect on human consciousness has been its coupling with the life sciences in disciplines such as genetic engineering and psychoneuroimmunology. The new perceptions that made these collaborations possible also created space for new assumptions about the nature of mind itself. The stage was set for a revolution in consciousness whose impact continues to reverberate. Willis Harmon, one of the most insightful futurists of our time, has indicated several of the most significant conversions in thought that have arisen due to this sweeping change in the concept of human consciousness:

- It has modified and qualified the way we interpret science.
- It has altered considerably how we think about health care.
- It has drastically changed our concepts of education.
- It has provoked major shifts in perception in the worlds of business and finance.
- It has been influential in the process of divesting the idea of war of legitimacy and validity in the world of sane men and women.
- It has influenced a call for a total rethinking of plans for achieving and maintaining national and global security.[6]

These shifts in worldview tie into the rhythm of the 1960s, when new ideas took hold, social mores changed, and a general sense of powerlessness and restlessness began to permeate our culture. People strove to identify themselves anew, and an avid interest in seeking out the fullness of their abilities opened the door to the human potential movement, led by Abraham H. Maslow,[7] Viktor Frankl,[8] Robert Assagioli,[9] and others. Until his death, Maslow led the way with his recognition that not only were we subject to a hierarchy of basic needs for survival, security, and self-esteem, but beyond these each individual had growth needs, actually metaneeds, for self-expression in work that was significant and self-rewarding. He contended that these metaneeds had to be acknowledged before transcendent needs could emerge, and that if metaneeds were ignored, "the soul would surely die."

The human potential movement gave birth to a revolutionary redefinition of man and woman. At the time, psychohistorian Marilyn Ferguson wrote, "A leaderless but powerful network is working to bring about radical change in the U.S. Its members have broken with certain key elements of Western thought, and they may even have broken continuity with history."[10] These persons she called the Aquarian Conspirators, and your author is proud to have been among them by virtue of the reformational effect Therapeutic Touch has had in the fields of education and health.

As this surge of new conceptual frameworks took on stability and depth, it loosened the rigidity of centuries-old models of scientific thought, opening a floodgate of inquiry into the nature of mind and its consequent effects on human functions. By the 1990s there had been a decided shift from the innate ideas and figures of speech of physics to the models and metaphors of biology, so that now even the "new physics" thought in biological terms and analogues. The terms microintensive, adaptive, holistic, computer virus, mouse, and feedback are a few of the bioterms now in common usage that suggest a more life-affirmative context.[11]

Intensive studies on extended human abilities (also known as ESP) by life science/high-tech research teams finally led to an acknowledged acceptance that human consciousness was not limited by the physical functions of the brain. Non-scientists, too, explored transcultural teachings on meditation, visualization, and imagery. Their personal experiences of non-local and non-temporal being touched off popular and formal inquiry into the farther reaches of the human mind. In the 1990s, C. J. S. Clark of Southampton University in England used quantum logic to develop a new model of human consciousness. The basic assumption underlying his hypothesis was that consciousness is non-local; that is, human consciousness does not belong to or exist only in a particular locality (e.g., in the brain or the immediate surroundings), but can transcend the known parameters of space and time and affect living organisms, objects and events.

Clark was not alone in his supposition. Nobel Prize–winning physicist Brian Josephson also proposed a non-local model for human consciousness. The idea took hold on a more popular scale as well, with new interest in the effects of prayer, healing at a distance, and mind-to-mind healing (such as may occur in the advanced practices of Therapeutic Touch). By October of 1999, the haute elite of conservative medical journals, the *Archives of Internal Medicine*, was printing evidence that demonstrated the benefit of prayer for the sick, even when the patients were unaware of supplication being offered for their welfare.

As it turns out, the model of a non-local consciousness is serving as a useful connector between the enlightened views of the late twentieth century and the intensive researches into human consciousness of the twenty-first century. Today's research is eliciting insights into the bewildering multiplicity of problems we are now facing, such as the profound and long-lasting stresses to which all contemporaries of our time are subjected; the far-reaching, demeaning influences that encourage us to casually regard the pandemic of substance abuse currently sweeping the most remote outposts of civilization; and the effects these acts of self-indulgence are having on our individual and societal consciousness.

An important element of these studies—to the surprise of those whose notions of reality are still locked into three-dimensional, Cartesian concepts—is the essential holism that studies of the many physical, psychological, mental, and spiritual aspects of human consciousness are demonstrating in the fields of psychoneuroimmunology, depth psychology, and conscious exposure to spiritual experiences, among other avant-garde researches. As awesome potentials of human consciousness are being realized, it is recognized that human consciousness has many states, levels, or perspectives. A short list would include the following levels of consciousness:

awareness	contemplation
sensation	spiritual feeling
imagination	ecstasy
passion	emotion
intellect	intention

At this time, study of these levels of consciousness remains within a penumbra of caution that delineates Western appreciation of the profundity of human consciousness. In other parts of the world, however, the dynamics of human consciousness have long been considered an appropriate subject for investigation. The renowned yogi, teacher, and philosopher Swami Rama states authoritatively that all of our information about the numerous energies that form the complex patternings of consciousness "come[s] to us from the experiences imparted through oral tradition by great meditation masters who have followed a path of self-awareness."[12]

Within the Eastern frame of reference, consciousness and life are reflective of each other. Consciousness turned inward is what we call life; life turned outward, we call consciousness. The major Eastern belief systems, Hinduism and Buddhism, hold that the consciousness of each person is essentially unitary and complete, no matter how separate and different from each other they may appear at a particular moment. It is their unitary nature that links us together and gives each person open access to the consciousness of the great Teachers. It is because each of us is an essential part of that consciousness that we all react similarly under similar conditions. The goal of one's own self-consciousness is to realize that this unitary nature of all people is fundamentally derived from a single source; therefore, it is through this primal source of consciousness that we can validly claim kinship to each other as brother or sister; even though we are born into individual families, we are kin to one another in this sense.

CHAPTER VII: THE CHAKRA COMPLEX: SOURCE OF CONSCIOUSNESS

Transforming Universal Energies into Human Consciousness

Probably the most poignant question humans have asked themselves through the ages (right behind the essential inquiry asked by every generation: Who am I?) is: How do I know?

At our present stage of understanding, what we envision is a universe in which physical matter is in constant, subtle movement or flow. These energetic, patterned swirls are impressively responsive to mindful direction. This mindful control, accomplished with intentionality, operates with increasing sensitivity throughout the levels of human consciousness. In the process, it can translate these energies of physical matter to the emotional, mental, and spiritual levels of consciousness. At the spiritual level, these energies appear to become immersed in the dynamics of another dimension that operates beyond our ken.

In Sanskrit, the fields that impose patterns upon these various human energetic flows are called chakras. In the Upanishads, India's most ancient and authoritative literature, chakras are referred to as *caitanya*, centers of consciousness. Each chakra can be said to have a domain or set that is related to distinct kinds of consciousness in the individual being. At the personal level, chakras derive as localizations, or local fields, from seed chakras or universal fields of consciousness that act on a cosmic scale. The relationship between these universal and personal fields is analogous to that of a universal electromagnetic field that affects the entire cosmos, and a localized electromagnetic field that helps to energize the individual living being. While each field of consciousness is important in itself, its power comes through its unique relationship to all the other chakras in the complex. And, as with physical fields, interaction across these consciousnesses or mind fields is instantaneous.

Within the Eastern conceptual frame discussed in the previous chapter, it is consciousness that provides the power (or drive) for will, knowledge, and action in daily life. These become effectual and functional through the chakras, as these non-material centers transform, channel, and conduct our vital energies. These energies, called prana, respond to universal rhythms (e.g., biorhythms) and are transported under their impetus. It is this constant flow of prana that energizes and maintains the physical body in health, and refurbishes and invigorates it in times of recuperation. Chakras, therefore, are the essential link to consciousness and its various levels of awareness.

Attributes of the Chakras

There are seven of these fields of consciousness that are of particular interest in Therapeutic Touch. All of them are nonphysical; that is, they are foci of force within the human-energy field. It was C. G. Jung who first brought the concept of the chakras clearly to the attention of Western thought, saying: "We are studying not just consciousness, but the totality of the psyche . . . [the chakra systems] are intuitions about the psyche as a whole, about its various conditions and possibilities."[1]

Dora Kunz has reported three distinct attributes of the chakras at the level of the vital-energy field:

1. At their cores, the vortices concentrate the energy within the field by drawing in the energy in a "tight flow." They then distribute it along the outer edge of the arms or petals of the chakras in successively widening spirals.
2. Each chakra has a distinctive and unique geometric structure.
3. There is a constant rhythmic flow of patterned colors within each chakra.[2]

Through the "texture" and rhythmic movement of their subtle energy patterns, and the size and intensity of the colors of their whirling arms, chakras express both the quality of consciousness and the extent of the individual's personal development.

The principal forces flowing through the chakras include:

1. Kundalini, the creative fire that is said to reach up from the earth's depths to dynamically engage the self-disciplined individual's chakra complex in a manner so inexplicable in common terms that its study has been guarded through the ages
2. The life force, the ability to initiate and set in motion the life process
3. Prana or vitality, the ability to organize and sustain life

It is essentially in the central nervous system, the autonomic nervous system, and the endocrine system that these nonphysical fields of different kinds of consciousness have their physical effects.

"Opening" the Chakras

It should be noted that the TT therapist uses her own chakras either to support, to amplify, or to be a model for the healee's chakras' functions, or to create a condition or ambience for certain kinds of healing to take place. These therapeutic strategies occur quite naturally;

the very expression of empathy for someone in need is itself a function of the heart chakra.

Such use of the chakras in Therapeutic Touch is in accordance with authoritative ancient teachings and our own contemporary experience. The idea that a healer can "open" or "close" a healee's chakras (a misconception that has arisen in certain contemporary healing modalities) is *not* supported by these sources.

The literature specifically states that "opening" the chakra can only be done by the individual himself or herself, and describes the method in explicit detail. The person is taught to go within to his or her interior self, following each felt pulsation of his or her body toward that consciousness within, "first as a point of focus and then as a thread leading inward." The person then closes the "outward flow" by concentrating only on the flow of consciousness within. This opening of the chakra leads to what we know as a relaxation response after the flow is reversed, and can result in a sense of unified experience.

As the experience takes place under the supervision of a teacher, there probably is much more about the exercise that is conveyed orally. However, even though the above description is minimal, one can see that what has been claimed recently to be the "opening" of a healee's chakra during a healing session is a misnomer as well as a misinterpretation of its purpose.

It is our position that except under unusual circumstances, no one but the individual him- or herself can initiate and sustain his or her own chakra functions; claims to the contrary are groundless and can be damaging to the healee. Also, the technique of "opening" one's chakra is actually a skill sought after by the committed student of yoga. In its largest sense, it is not primarily a therapeutic technique.

The Root Chakra

The seven chakras that relate to Therapeutic Touch are in reference to the upright physical body. It is said that from the base of the spine to the top of the head, there is an increasing differentiation in the vibratory rate and in the functioning of the chakras, resulting in more subtle awarenesses being brought into consciousness as one gains successive control of the higher chakras. The *muladhara* (*mula* = root) is the lowermost chakra. It is "located" at the base of the spinal column between the anal orifice and the genital organs, at the site of the sacrococcygeal plexus. (Of course, in discussing these sites, it is realized that though one speaks of a physical "location," the chakras are nonphysical. They project through the three-dimensionality of the physical sites, and are not bound by physical barriers such as biological membranes.)

It is in the muladhara that kundalini, which is concerned with multifaceted creative forces, "sleeps" in a latent state, blocking access to the prime nadis (the channels of conduits for pranic currents, as mentioned above).

The root chakra itself is said to be related to instinctual forces deep in the being of the Earth, to the cohesive power of matter, to inertia, and to the sense of smell. It is also considered to be the basis for the present evolutionary thrust on Earth. On a basis of knowledge that is difficult for us, embedded as we are in Western civilization, to understand, its sphere of influence is said to be concerned with ripening karma, rigidity of thought, and routine activity.

In Therapeutic Touch, it is mainly the so-called upper chakras (from the heart to the crown) that are used. The root chakra site is primarily accessed to transmit high energy to healees who have fractures, venous stasis ulcers, or other imbalances of the lower limbs, who are paraplegic or who have severe lower gastrointestinal problems.

The Spleen Chakra

Next is the spleen chakra, the chakra directly related to the channeling of pranic flows within the individual. In the ancient literature, the spleen chakra is said to not be a chakra; nevertheless, it is more significant than a term such as "quasi-chakra" might indicate. The spleen chakra specializes, subdivides, and distributes the vitality that it transforms from the sun's energies. This means that prana, a complex of seven solar energies (of which we are able to assimilate only five), breaks down at the spleen chakra into its component parts, which recombine in various ways to vitalize the organs of the body.

This pranic flow seems to be intrinsic to what we call the oxygen molecule; from a Western physiologic point of view, the spleen is concerned with the regeneration of red blood cells. Red blood cells contain the porphyrin hemoglobin, a respiratory pigment that distributes necessary oxygen to the tissues. Because it is the oxidative process that vitalizes the human body (i.e., human metabolism is mainly oxidative), the functions assigned to the spleen chakra seem rational from a Western point of view.

The functioning of the spleen chakra is disturbed by fatigue, sickness, and extreme old age, and therefore in Therapeutic Touch we work to support the efficiency of this chakra.

The Solar Plexus Chakra

Ascending up the spinal column, the next chakra is called the *manipura* or the *nabhi* chakra, and is situated in the lumbar region at the level of the umbilicus (*nabhi* = umbilicus). It is said to be related to fire and the sun, and affects menstrual flow, breath, sight, and the transmutation of organic substances into the psychic energies of the lower, more sensuous emotions. From the current psychosomatic medical point of view, it is located in the abdominal area in which one gets peptic ulcers or "stomachaches" as a result of pent-up anger, anxiety, and other intense emotional states.

The Heart Chakra

Continuing up the spinal cord, one finds the *anahata*, the heart chakra, which is in the region of the heart. It is considered the seat of prana, the life energy, and of the *jivatman*, the individual soul. Its functions are related to air, touch, the motor forces of the body, and the blood system. In Western terms, it regulates respiration, the heart, and the blood through the sympathetic nervous system.

The anahata governs the plane of human realization. In Therapeutic Touch it is used primarily to protect or calm the healee or to sedate a condition. Most importantly, since the heart chakra region includes the site of the thymus, it can be used to stimulate the immune system.

The Throat Chakra

The *vishuddha* chakra is at the site of the laryngeal and pharyngeal plexes, at the junction of the spinal column and the medulla oblongata, where the base of the head meets the upper vertebrae of the spinal column. It is said to be the seat of the breath that carries the soul to the head in the saint's ecstatic state, known in Sanskrit as *samadhi*. It is the site of conscious transformation of prana through mantric power, which is accessed through knowledge of the functions of controlled sound, such as mantras, chanting, and some forms of breath control. It is also related to the skin, which is highly sensitive to vibrations.

Physiologically, it should be noted that the site assigned to this chakra, the medulla oblongata, contains the vital centers of the body. This chakra site is primarily used to access the healee's inner self when there is need for heroic efforts, or the patient is in panic or frightened.

The Brow Chakra

The *ajna* chakra (ajna = order, command) is between the eyebrows, in the cavernous plexus. It is the seat of the cognitive and subtle senses, and the TT therapist uses it during the assessment phase to visualize sites of imbalance in the healee's vital-energy field.

The Crown Chakra

The *sahasrara* (sahasra = thousand), also called the *brahmarandhra* chakra, is at the top of the head. It is symbolically represented by a thousand-petaled lotus. It is associated with the pituitary-pineal axis and is concerned with volition, acts of will, and altruism. The crown chakra is our gateway to consciousness and our entry point to states beyond our ordinary consciousness. In Therapeutic Touch, the therapist uses the crown chakra primarily to establish mind-to-mind contact with the healee for purposes of healing or communication.

The Pranic Flow

The incessant upwelling of pranic flow is individually tailored to each person by his or her chakras. This pranic flow makes itself known through its masterly organization of the life process, and is perceived through the physical principles that underlie bodily movement, change, and transformation. The pranic flow also produces the states of consciousness that we know as attitudes, behaviors, and awarenesses. When we apply both of these concepts, we find that we are matrices of energetic flows from the heaviest matter of our bodies, i.e., bones, to the lightest of thoughts.

The Nadis

In the literature it is said that there are patterned lines of force (nadis) that carry the pranic flow within each individual's vital-energy field. The three major nadis, *sushumna*, *ida*, and *pingala*, are said to "slumber" within the muladhara (root) chakra until aroused by the awakening of the primal creative force (kundalini) latent in all warm-blooded creatures. Sushumna is described as running up a lumen that is hollowed through the center of the spinal cord; Avalon reports it to be the remnants of the hollow tube from which the spinal cord and the brain developed in the embryo.[3] The companions of sushumna, the ida and the pingala, are said to be related to lunar and solar energies respectively. These two nadis spiral in opposite directions, entwining sushumna as they all ascend the field counterpart of the spinal cord.

In the course of this ascent, the elongated stem of each of the chakras becomes embedded in the flow of the major nadi, sushumna, and the three nadis (sushumna together with ida and pingala) collectively become the channel for the fundamental life force (prana) of the individual. In some manner not easily described, the unaroused or latent kundalini, rising naturally with the nadis, meets this basic life force. These forces are said to revolve in opposite directions, thereby increasing the pressure within the system and giving rise to what we would call personal magnetism. The ida and pingala weave through the sympathetic nervous system as it runs down each side of the spinal cord. These ganglia of the sympathetic nervous system fan out from the spinal cord as they proceed downward from the base of the brain to the coccyx. Finally, sushumna, ida and pingala join in a "triple knot" at the site of the ajna (brow) chakra. The three nadis then exit the human field through the field counterpart of the nostrils.

Energy Flows and the TT Assessment

Awareness of the chakras and their functions allows us to better visualize and understand the TT process. Let us examine Therapeutic Touch in the context of the energy flow dynamic explained above.

Compassion for someone who is in need brings one to the TT interaction, but once that determination is made, the centering of the therapist's consciousness is primary. That critical shift in consciousness is a fundamental first step and is maintained throughout the TT session, the other phases of the TT process proceeding even while the focus of attention remains in that centered state.

The therapist remains relaxed but alert, and is open and sensitive to the needs of the healee. She is in an attitude best described as "deep listening," searching for subtle cues to the healee's condition. Concomitantly, the therapist often senses that the heart chakra is actively engaged in the healing interaction. There is a lessening or disappearance of environmental noise, a sense of stillness as one goes about this inner work in a mindful manner.

The TT therapist first conducts an assessment of the healee's vital-energy field in an attempt to determine where the subtle energy imbalances are. This involves a deep focus of the therapist's attention, akin to the act of attentive listening, and the use of her hand chakras as telereceptors. As the hand chakras explore the healee's vital-energy field, the TT therapist becomes aware of deviations from the basic rhythmicity and subtle energy patterns of the healee's vital energy field. These indicators of vital-energy field imbalance are called cues; however, since the English language does not have adequate descriptors of those subtle energy states, the terms in which these cues are expressed are recognized to be but metaphors for the nonphysical dynamics of the healee's vital-energy field. These include:

- Breaks in energy flow
- Vital-energy deficits or vital-energy hyperactivity
- A sense of pressure or fullness
- Congestion, sluggishness, or blockages of vital-energy flow
- Differences in temperature
- A sense of tingling or slight electric shocks
- True intuitions or insights

In order to rebalance the healee's vital-energy field, the TT therapist works to reinforce and strengthen the pranic flow within the nadis. From time to time she reassesses the healee's vital-energy field to determine whether his or her present needs for rebalance and integration have been met, and also to figure out what should be done next. As the cues in the healee's field are rebalanced and the intensity of the flow of information fades, the therapist recognizes another characteristic shift in consciousness taking place within herself. This modification of awareness signals to her that the healee's vital-energy field has been rebalanced, and she completes the TT session.

The effect on the therapist is vitalizing, and feelings of tiredness are rare. Often the healer has an altered perspective of the healee, realizing that she now understands the healee more deeply than before the TT session began. There may be fuller and deeper insights into how she can most comprehensively help the healee.

History of the Chakras

There is a strong agreement among healers concerning the various chakras they use under different conditions of healing. Whether the healers have heard about the concept of chakras or not, their experiential descriptions of specific chakras closely agrees with the functions ascribed to them in the ancient literature of India.[4] Explicit information about the chakras goes back about four thousand years in India, in sacred Hindu texts called the Upanishads. However, ancient knowledge of the chakras is not limited to India. It can be found, for example, in the Popul Vuh, a very ancient text known throughout Central America. The chakras were discovered by geographically distant peoples, such as the Sufi in the Near East, the Apache, Hopi, and Zuni in the southwestern United States, and the Hunas in Hawaii.

In the Upanishads, chakras are described as being complex, whirling, rhythmical energy flows that cohere in multidimensional space. At their nexus, the streaming of universal energies from the cosmos and psychic energies from the individual growing fetus intersect to form a gyrating vortex out of which the form and organization of the individual emerges. The result of this convergence is the materialization of "bodily qualities"; that is, of physiological functions of the fetus. The Upanishads say that later in life these bodily qualities can in some way be transmuted into psychic forces again.

In this way, the chakras are natural components of each individual's human-energy field, whose major function is to transform universal energies into those that define us as human beings (e.g., the awareness of objects, events, interactions, and relationships in a context that we call the mind).

Chakras and Consciousness

To view human consciousness from the perspective of ancient Eastern cultures—for Western culture is, admittedly, only beginning to understand the farther reaches of the mind—one must subvert an accepted Western perspective about human growth and development.

While Western tradition treats the physical body as the basis and goal of human manifestation, the writings of the Eastern schools of thought agree that the beginnings of human consciousness and its embodiment in physical matter occur at the transpersonal rather than the physical level. In a manner said to be dependent on the individual's karma (a profound and complex concept based upon an assumption of reincarnation as fact; a force generated by one's actions in previous lifetimes that determines the nature of one's current life), appropriate human energies are attracted to a non-physical site in space and time. This site is the focus for the primal energies that orchestrate the growth and development of a human fetus and lay the primal foundations of the chakras.

Although Western models assume that the brain is the exclusive seat of consciousness, yogic experience has demonstrated that our brain-consciousness is only one among a number of possible forms of consciousness and that these, according to their function and nature, can be localized or centered in various organs of the body.[5] From the chakras associated with these organs, secondary streams of psychic force radiate, much as the spokes of a wheel radiate from the wheel's hub. It is in this manner that psychic forces penetrate the physical functions of the growing fetus. The relationship between the broader chakra-consciousness and the psychic force lays the foundation for the psychosomatic functioning of the growing child and its several levels of consciousness: the spiritual domain, the realm of the inner self, the conceptual field, the psychodynamic field, and the physical field.

Although the term "levels of consciousness" has been given to the fields listed above, to properly get a perspective of the dynamic reach of human consciousness, it must be recognized that these fields are not isolated from one another; they interpenetrate, interrelate, and intercommunicate constantly.

In addition, to maintain an appropriate frame of reference it is important to remember that these fields are but local representatives of universal fields of consciousness. The universal fields' primary focus is relative to the concerns of the cosmos, whereas each localized field functions in its own differentiated space and has qualities and expressions that are specific to the individual person, as well as universal qualities of humanness that can be thought of as archetypes.

Govinda's Concept of Relationship and Universal and Local Fields of Consciousness

Lama A. Govinda writes that the relationship between the universal fields of consciousness and the localized fields of the individual fall into three general realms or zones.[6]

The first or upper domain is concerned with the universal, eternal laws of the cosmos. Knowledge and function within this perspective operate beyond the confines of ordinary time and space. We are aware of this domain primarily through the brow (ajna) and throat (vishuddha) chakras.

The conditions of the second zone allow the individual to realize "qualities of territorial existence and cosmic relationship that become conscious in the human soul as an ever-present and deeply felt reality." This comprehension of at-one-ment is related to the heart (anahata) chakra.

The third or lower zone is concerned with terrestrial, earthbound forces that ground physical experience. These are mediated through the solar plexus (manipura) chakra.

The Chakra Complex

Another way of comprehending chakras is to think of them as nonphysical two-way converter stations, where psychic forces and physiological function merge and permeate each other. They are transmitters sending messages, and transformers that reduce or increase energies in order to move between levels of consciousness. It is in this manner that they bring about a coordination of the psychodynamic, the conceptual, and the vital energies, regulating their individual frequencies. They dovetail neatly, engaging and integrating with each other to regulate their individual frequencies in a harmonic net of coordinated functions.

When a human is born as a conscious being, all the individual's chakras, comprising in the fullness of their expression a chakra complex or "thread of chakras," integrate their functions into a synchronized, unitive network that embodies the full range of consciousness of which humans beings are capable. It is in this way that each individual is intimately linked—"at one"—with the formidable forces of the universe and the natural world of our planet.

It is because of the potential we have to span this intelligent vastness that the human-energy field has been called "the silence" that hovers as potential, not actualized into materiality, just below the quantum mechanics of the physical universe. This conception suggests that an innate process for control ("the silence") is constantly aware of the interactions of the parts as they play against the dynamics of the whole. As scientists begin to probe deeply the inscrutable mysteries of the levels of human consciousness, they are realizing that there may be a way to integrate "the silence" by insightful use of the levels of human consciousness—and that a sensitive understanding of ancient teachings about the chakras is key to this comprehensive vision.

SUMMARY: THE CONCEPTUAL FRAME OF REFERENCE FOR THERAPEUTIC TOUCH

The core idea underlying the conceptual frame we visualize for Therapeutic Touch is based on universal principles of order. Within this context there is a multidimensional universal healing field whose sphere of influence promotes the reestablishment of a state of balance to living beings who are injured, debilitated, or ill. In Therapeutic Touch this is done through the rebalance of the healee's physiology, psychology, levels of consciousness, or state of spiritual equipoise.

The medium through which Therapeutic Touch operates is understood most readily through field theory. In the context of Therapeutic Touch, the operant medium is the vital-energy field and the active force is prana, empowered by the TT therapist's alignment with her inner self in compassion for those in need.

The fundamental factors in the practice of Therapeutic Touch are the TT therapist's knowledgeable use of her chakras with intentionality, an empathic recognition of the limitations and aberrations of the healee's imbalanced vital systems, and, most importantly, the therapist's uncoerced extension of permission to her own inner self to integrate with her physical, emotional, cognitive, and spiritual levels of consciousness in the compassionate interest of helping or healing someone in need.

Therapeutic Touch and Western Science: Conceptual Similarities

In comparing the concepts around which Krieger and Kunz developed Therapeutic Touch[1] to those that are embedded in Western science, it should be made clear that the comparability is valid and not merely a coincidence of terminology, and that there is a high correlation between the contexts in which concepts from these frames of reference are used. Several significant mind connections bridging Therapeutic Touch with Western science are listed below, and it is suggested that the teacher of Therapeutic Touch use them to start open, ongoing dialogue with her students on the implications of the conceptual framework of Therapeutic Touch.

A prime belief shared by Therapeutic Touch and Western science—probably the most significant breakthrough to the beginning understanding of the human potential—is that consciousness does not reside only in the brain. This concept is important because it laid a formal basis for the serious acceptance of self-conscious mind as a human function that transcends the biological level of physiology and biochemistry. It was this acceptance that man is more than his physical body that made possible the radical leap to the alternative realities that fired the imagination during the last quarter of the twentieth century and set the tone for the new millennium.[2] That acknowledgment was also conducive to the acceptance of Therapeutic Touch by the Establishment of that time, Western science.

A second agreement is the realization that the human being not only possesses multiple seats of consciousness, but also has access to several levels of consciousness, which are well integrated in the healthy individual. Much of this understanding was achieved through the frontier neurosurgery of Sperry,[3] Penfield,[4] and others, beginning in the mid-twentieth century. Sperry's major contribution was that in dissecting the brain longitudinally, he found that each hemisphere of the brain had different, specific functions. Penfield is noted particularly for his finding during surgery that pressure on various regions of the brain stimulated different functions; some of which concerned the detailed awareness of happenings at a distance that were verified later, and others arising out of memories and flashbacks of an earlier time.

A third area of agreement has come more recently to Western science through studies on quantum theory. These findings recognize that consciousness is as important and independent a state as is energy, matter, or information.

Fourth, field theory, as defined by psychology and the social sciences in general, forms a compatible basis of understanding between the conceptual framework of Therapeutic Touch and that of Western science.

Finally, much of the TT process occurs in the healee's vital-energy field, which is non-physical and therefore cannot be seen by most people. Because of this inability to see the vital-energy field with the eyes, in the early days of Therapeutic Touch those who were not familiar with the new physics doubted our credibility. However, it is now generally known: modern science accepts that one's inability to see an entity does not mean the entity is not real. In Western science, the notion arose out of a progression of ideas that were nested in quantum theory, but it has found a pragmatic base in the nanotechnology of our time, which has opened the world of the very fine structures to modern industry.

Therapeutic Touch and Western Science: Conceptual Differences

An outstanding and fundamental difference in the conceptual frames of Western science and Therapeutic Touch rests on the Western view of human functioning. Western science sees the physiochemical and psychological faculties as the goal and basis for human evolution, but in the TT view, it is the spiritual aspects of one's being that underlie the human physical evolution and emotional aspirations toward "becoming fully human."

A second critical difference is born out of the mature TT therapist's experiential knowledge that consciousness can extend to supraphysical levels. The closest correspondence in Western science is the acknowledgment of the validity of intuitive hypotheses.

A third area of major difference lies in the acceptance in Therapeutic Touch that there are non-physical and therefore non-measurable complexes that form the centers of consciousness in the human being. The majority opinion in Western science is that consciousness arises in the brain. In Therapeutic Touch, the conceptualization of these centers, called chakras, rests on Eastern theoretical and philosophical subjective findings. These findings have come down through millennia of a consistent tradition of personal and committed inquiry, directed toward the intense study of one's own consciousness for the purpose of understanding the individual's relation to the universe.

An essential experiential realization of the practicing TT therapist that fundamentally differs from accepted precepts of Western science is that spiritual aspects of oneself (the inner self) inform and guide the physical self, and may direct the conditions of one's life.

Topics for Further Clarification

Conceptual frames of Therapeutic Touch are substantially based on experiential knowledge gained during the healer-healee interaction during the TT process. However, even the therapist in action may have a limited understanding of the details and the implications of the process. Further conceptualization of the TT process is needed in order to make clear precisely how the TT therapist "connects" or makes deep healing contact with the healee.

The continuing development of concepts associated with Therapeutic Touch will also help to explain certain phenomena that have been known to surround the TT session. For example, the healee may feel a sense of prior knowledge about the therapeutic relationship with the therapist. Or the therapist may feel after the TT interaction that the healee, perhaps a stranger before the TT session, is now deeply known or profoundly familiar. This state of "knowing" the healee is difficult to explain objectively, and the details of how the perception is occurring may seem a mystery. Since "feeling" something but not understanding it lessens the TT therapist's confidence in how she perceives the healee's condition, this issue also needs clarification.

Further conceptualization of Therapeutic Touch will also be of use to therapists who wish to integrate their practice with particular physical or spiritual theories. Those who subscribe to the idea that death is but one of many transitions in the continuum of living consciousness, for example, may wonder how the karmic derivation of the healee's state of consciousness affects his state of health and his ability to be healed. Or a therapist may ponder the enigma of what the individual is as a "total being."

Finally, the TT process relies heavily on the TT therapist's personal experiential knowledge, gained during the process itself. Although the intervention is individual, the majority of TT therapists agree on what clinical practices do and do not constitute the so-called Krieger-Kunz method of Therapeutic Touch. Nevertheless, there is need for a generally accepted objective standard by which society can gauge the validity of therapists' personal interpretations of what is happening during the TT interaction. Such a standard should include considerations of how long the healing should last, whether healing has in fact occurred, and how thorough it has been.

PART II

THE EXPERIENTIAL FRAME OF REFERENCE FOR THERAPEUTIC TOUCH

THE PERSONAL EXPERIENCE OF THERAPEUTIC TOUCH: AN OVERVIEW

Clearly defined phases mark the therapist's progression into the TT process. First, of course, is compassion, the felt need to heal. As she becomes involved with the TT process, the therapist feels an irresistible urge to make a channel of herself for healing energies and, in so doing, to be at one with her higher orders of self. In this act of centering her consciousness, the TT therapist begins to realize the existence of an implicate or enfolding order that underlies the more physical or unfolding order in the world.

As discussed in the previous section, it is within this implicate order that the concepts of a universal healing field and the human-energy field have relevance. Each individual arises into being as a localization of the human-energy field. A characteristic function of the human-energy field is the chakra complex, which gives rise to the spectrum of human consciousness in the individual person. One of these levels of consciousness, the vital-energy field, vitalizes and invigorates the human being. The vital-energy field is energized by subtle energies that respond to Therapeutic Touch.

As the therapist makes Therapeutic Touch an intrinsic part of her lifestyle, she may gradually become aware of the availability of a profound underlying level of consciousness, reaching beyond simple survival and personal needs. As that recognition deepens, the TT therapist begins to realize the intimate relationship of this level of consciousness to her inner self. Should the TT therapist be willing to align herself with that level of consciousness, her inner self will manifest itself more discernably in her daily life. It is then that acts of intentionality, grounded in clear visualizations and mobilized as powerful living forces, will present themselves during the healing process.

Such intervention will occur most vividly during the rebalancing phase of Therapeutic Touch, as the TT therapist, with specific and informed intentionality, repatterns the flow of vital energies by directing, modulating, or unruffling the vital-energy field to bring the field into balance.

It is in this response to an imperative that originates beyond the level of her ego needs that the TT therapist becomes a human support system, helping the healee, whose vital-energy fields have been debilitated, traumatized, or sickened, to regain control of his immune system by the rebalancing of his vital-energy field.

During this engagement, both healer and healee are exposed to potent experiences that may help them transcend the limitations of their individual personalities. Such quickened transformations seem to go beyond the usual constraints of space and time, so that the healer—who in compassion reached out to others—now may become a teacher; and the healee—who originally was impelled by a desire to be healed, i.e., to be made whole—may now be motivated to become a healer to help those who are in need now.

It is during such transforming experiences that one begins to realize that the Therapeutic Touch interaction is not so much a technique as it is an inward journey, an inner growth experience birthed in compassion and matured in conscious communion with one's inner self. In the concluding chapters we will examine these constructs and concepts again in the perspective of the experiences they bring forth.

CHAPTER VIII: THE POWER OF COMPASSION: A POTENT LINK TO THE INNER SELF

Therapeutic Touch: A Technique of Compassion

Why do people want to help or heal others in need? Usually it is sympathy, empathy, or compassion that motivates the healer in this most humane of all human acts. There is no doubt that it is from the welling up of compassion within the therapist that Therapeutic Touch initially draws its power. In fact, after Eliade's definition of shamanism as "a technique of ecstasy,"[1] I have often thought of Therapeutic Touch as a technique of compassion.

Compassion has the power to change our perceptions, so that the person dying alone is no longer a pariah to be shunned; the person in pain is no longer a nuisance; the person festering in illness is no longer ignored, but is helped.

The Feminine Principle and Compassion

Compassion is often mistakenly thought to be a quality of women rather than men. Compassion is one of the qualities of the feminine principle, a natural quality inherent in both males and females. (Other distinctive attributes of the feminine principle are support, nurturing, love, and aesthetics, all qualities incorporated in the healing act.) Compassion, then, can be common to all people.

Compassion as Silent Mantra

Compassion has the power to urge us to work beyond our usual capacities to meet the needs of another. In a sense, compassion is a magical power, for under its impress the impossible becomes practical. When confronted with other strong emotions, such as anger or jealousy, compassion releases a cascade of hormones and other biochemicals, changing the mood of the TT therapist and helping her project healing energies of a wondrous nature and an awesome power to the healee.

However, for this to happen most effectively, the setting must be prepared. In Therapeutic Touch, this occurs as the therapist gives her full concentration to the task at hand, bringing a sense of peace and equanimity to the healing milieu. It is in the still ambience of unconditional love that the call within can reach the level required to effectively project healing energy. It is as if the welling up of compassion from the depths of the individual resulted in an inner sound or vibration.

This inward channeling of vital energy seems similar to what is called *shabda*, a "silent" mantra that can set into motion one's own chakras. Since compassion is a necessary state of consciousness for the TT therapist, in this case it is the heart chakra that is energized. It is interesting to note that in the Upanishads it is stated that there is a subtle energy connection between the heart chakra and the sense of touch, a physical—as well as subtle—function that is central to the Therapeutic Touch interaction. It is said that, like a mantra, this "inflowing energy" or inner-directed effort must be realized in the depths of the human heart in order to be transformed into vibrant being.

Without compassion, whatever else one may do, it is not Therapeutic Touch.

Compassion as Ally

As true compassion works through the deep places of the heart chakra, it becomes an ally in the TT therapist's healing work. The irrepressible surge of love as compassion floods through one's being, coupled with the associated upwelling of aspiration and identification that seems to activate the throat chakra, awaken one's intuitive capacity, and further sensitize one to the promptings of the inner self. The experience of this integrated upsurge gives rise to the supposition that the coherent force of this released energy breaks through restricting patterns of conditioned and habitual vital-energy flows to change their level of resonance. It is at this time that the TT therapist may experience a sense of the order and unity that underlie the healing process and, in identifying with that epiphanic moment, become a channel for healing.

As one studies the world literature on healing, the impression deepens that it is compassion and a sense of order that underlie the aspirations of all healers; it is at this level of functions of the higher orders of self that all healers meet. The common ground for this insight appears to be a prior acceptance of the personal vulnerability that accompanies the conscious demonstration of compassion, of the responsibility for intervening in another's life, and of the role of significant change agent for the healee. These, it seems, are the conditions, or permissions, one must accede to in order to pass through the gateway as healer.

This rite of passage imputes a mysterious power to the act of compassion. For instance, the act of compassion gives access to new states of awareness. For one moment, perhaps, one senses the healee and the circumstances differently. However, for that moment there is no denying the reality of that different perception, and, as an act of interiority, the TT therapist consciously responds to the dictates of this compassionate urge.

Under these circumstances, Therapeutic Touch can become the source of a self-realization of the power of compassionate relationship. For example, you may see a dirty drunk lying in a gutter. Something surges up spontaneously within you and spreads through your being when you see him. There is an instant identification that is unexplainable—transverbal, at best—and you are at his side, helping, in that instant. You are pulled there by an insistent force you may not understand, but nonetheless you note a flicker of recognition somewhere in your being. With a glance, an unspoken message has passed between the two of you. This mind-to-mind communication may result in unexpected cooperation from the alcoholic, a sigh of relaxation or relief—most often, he releases himself to sudden, deep sleep. In that healing moment you realize that there is a beauty to the drunk that you hadn't noticed before, and out of some deep place a confirmation arises that you are beautiful, too.

> Someday, after we have mastered the winds, the waves, the tides and gravity, we shall harness for God the energies of love. Then, for the second time in the history of the world, man will have discovered fire.
>
> —Pierre Teilhard de Chardin

CHAPTER IX: VENTURES IN TRANSPERSONAL HEALING: LIVING WITH MULTIPLE REALITIES

Overview

With a view to presenting a wide variety of teaching styles in Part III, this chapter will consider how the compassionate practice of Therapeutic Touch can offer an opportunity to operate at high levels of function to quicken and actualize natural potentials of being. This level of function is called the transpersonal because it allows one a perspective that is beyond the usual view of reality: a non-ordinary perception, a context for one's actions that is light years beyond the cause-effect impressions of the world we have come to take for granted as being a part of a single true reality.

To facilitate entrance to this realm, several variables that are of significance to the TT experience and set the stage for a natural emergence of the transpersonal will be explored. To help us recognize our inborn connection with our own inner being, a variety of introspective experiences will be suggested. We will examine common verbal expressions that are provoked by engagement in inner work, and focus in on the reality of the transpersonal itself.

Dual Aspects of Therapeutic Touch

As one watches a Therapeutic Touch session, it is difficult to realize that an experience that appears to be so direct and simple could call out significant shifts in consciousness in the therapist so engaged. However, it rapidly becomes obvious to the involved therapist herself that Therapeutic Touch is not as simple as it seems. In fact, its complexity is as deep as the therapist's understanding.

One way of viewing the TT process is in terms of two general aspects. One aspect is concerned with the theoretical content, research, and clinical studies on the TT process itself. The second is concerned with the experiential knowledge that flows from the TT interaction and leads to a personal knowing.

When actually functioning, however, these two aspects are coupled with many other ongoing functions, the whole engagement working synergistically to foster rapid personal growth experiences. For instance, as the healer is feeling compassionate concern and helping the healee, the inner work that accompanies compassionate healing proceeds. In this act of interiorization, both the theoretical and the experiential aspects of Therapeutic Touch are more clearly defined. Moreover, as the TT therapist matures in the practice of Therapeutic Touch, the entire complex of these aspects dynamizes to expedite and enhance the subtle, evolving interior processing that sets the stage for the enactment of the transpersonal in her life.

To clarify what Therapeutic Touch is and what it encompasses, it is useful to consider some of the variables involved as they relate to the transpersonal, at which level some insight is provided about the dynamics of the TT process itself. (See Table I: Dual Aspects of Therapeutic Touch, p. 91).

The Transpersonal Nature of the TT Process

At this writing, almost thirty years have passed since Dora Kunz and I first developed Therapeutic Touch in the summer of 1972. Much of what we now take for granted wasn't known then. The act of healing and its rationale relied at that time on a religious frame of reference, and science could not find an adequate context for it. Therapeutic Touch challenged the religious traditions of healing in its most basic assumption by asserting that healing is a natural human potential that can be actualized under appropriate conditions, i.e., the healer is not an especially chosen person. Therapeutic Touch also challenged the scientific perspective in that it "worked" even though we still don't understand how subtle energy is transferred from the healer to the healee.

In developing Therapeutic Touch, Dora and I used every means at our disposal, although in hindsight it is difficult to recall the precise order of the process. Probably the first avenue of the development of TT arose out of our experiential knowledge as we began to realize that we could help people who were ill; however, much of that initial understanding stemmed from our many observations of expert healers at work. From those observations and our own experiences we began to develop our fundamental suppositions. This was not always a straightforward, logical process; flashes of insight helped us leap forward, even in the face of logic that might point in another direction. However, our saving grace has been that we have always been willing to test our notions before putting them forward.

It was out of this amalgam that we developed our theories about the TT process, most of which we've had the opportunity to test again over the years. If there was a handmaiden to witness the growth and development of Therapeutic Touch, it probably was clothed as an uncomplicated desire to help those in need. And I think it was the persistent force of that unswerving urge that helped us grasp the reality of the transpersonal infrastructure of the TT process.

From the beginning, the development of Therapeutic Touch was sponsored by universities, hospitals, and the health professions in the United States (and later in a large number of the countries of the

Table I. Dual Aspects of Therapeutic Touch

THEORETICAL KNOWLEDGE	EXPERIENTIAL KNOWLEDGE
Compassion as power uses the heart chakra.	Compassion is an ally on the path to the transpersonal.
The TT therapist can learn to understand the dynamics of the vital-energy field without the use of contact-touch.	A progression of experiential knowledge occurs as the TT therapist learns how to become sensitive to the hand chakras, for the hand chakras are intrinsically related to the heart chakra, and the heart chakra is significantly related to the crown chakra. In the final analysis, it is the relationship between all the chakras involved that we will call "the experience."
Intentionality is an exercise in identity, a recognition of one of the signature powers of self.	Directing the vital-energy on the out-breath in the act of intentionality focuses the precise depth and direction of the TT therapist's projection of the vital-energy flow
Chakras are centers of different kinds of consciousness.	Chakras are functionally related to each other. For example, visualization is one of the functions of the brow chakra. Used for this purpose, it also stimulates mind-to-mind ability via the crown chakra, which synergistically affects the entire chakra complex.
Centering is an act of interiority.	Centering, and staying on center throughout the TT process, is an empowering agent for healing and also informs the TT therapist about her connection to her inner self.
Cues are a way of getting information from the vital-energy field; this information can be transmitted to other TT therapists.	During the TT assessment, there is a pattern recognition of constellations of vital-energy flows, psychodynamic motifs, thought forms, and other expressions of consciousness.
The Deep Dee is a method of relating to the deep structures of mind.	The Deep Dee is an arena for inner work.

world). Having that academic and professional background forced us to develop the curriculum in a formal manner that spoke to the validity and reliability of the theoretical content. The theoretical content could be tested and graded; thus it was established that Therapeutic Touch was not only teachable but learnable. With the continuance of Therapeutic Touch established and the consequent development of standards of practice and evaluation tools, Therapeutic Touch became a pioneer in the entrance of optional therapies into the arena of formal higher education, life studies, and adult education.

It was within the strictures of this pioneering role that it was verified that the TT therapist explores and maps the vital-energy field of healees in a conscious, mindful manner, and that the results of the TT assessment can be communicated and verified by other TT therapists. This confirmation laid the groundwork for the acceptance of the TT assessment as a new means of social communication—and it was while the nonphysical communication aspect of Therapeutic Touch was being studied that the transpersonal qualities of the TT process itself were made clear and methods for teaching it in such fashion began to be realized.

With practice, the TT therapist comes to recognize that TT is an invitation to move into a larger reality; it is, in fact, a challenge to venture into the realm of the transpersonal.

Fundamental Theories of Transpersonal Psychology

Transpersonal psychology, as a study of man's normal but often crucial relationship with the realities of the spirit, arose to public view in the late 1960s. It directly challenged the continuing mechanistic and reductionist scientific viewpoint, which ignored claims of spiritual reality. The acceptance of this new dimension of psychology grew out of A. H. Maslow's studies of peak experiences,[1] Tart's study of deep states of consciousness,[2] the Greens' biofeedback study of yogis,[3] Grof's extensive studies of the human psyche using psychedelic drugs,[4] and Ken Wilbur's profound integration of the psychologies of East and West.[5] These brilliant intellectual outpourings came on the heels of the popular introduction into the United States of Indian, Tibetan, and Chinese philosophies reaching back thousands of years. Over time, the field of transpersonal psychology has thoroughly integrated these ancient and contemporary views into a synthesis of optimal states of consciousness that is in accord with the leading-edge thought of our time.

The transpersonal is an intensely private, often secret experience. It seems to arise out of the same domain as does intuition and, like intuition, is often nonverbal. Tart considers transpersonal experience to be part of the "super-consciousness."[6] Transpersonal perception allows access to multiple realities, but the validity of this inner vision rests on uncommon clarity in the interpretation of one's perceptions, which are bolstered by strong intuitive insights and a stable, healthy ego. There is growing evidence that, if these attributes can be nurtured, a transpersonal state of consciousness can be reached that provides entrée to inner sources of illumination and inspiration, which can be precious allies.

As Grof's extensive studies on transpersonal aspects of human consciousness have demonstrated, these experiences seem to occur beyond the limits of space and time as we know them and to "tap directly into sources of information that are clearly outside the conventionally defined range of the individual . . . and span an immense uninterrupted experiential continuum."[7] The range of experiences that have been reported is all but inconceivable. There are detailed accounts of ancestral experiences, racial and collective memories, historical events, and an entire spectrum of extrasensory experiences that Grof considers valid transpersonal happenings. He asserts that "transpersonal experiences have many strange characteristics that shatter the most fundamental assumptions of materialistic science and of the mechanistic worldview,"[8] and, indeed, accounts of the transpersonal are so far beyond the comprehension of the traditional scientific point of view of the past four hundred years as to clearly mark our present time as being "between parentheses," a time caught between changing paradigms.

Ken Wilbur speaks to the many dangers that threaten the clear understanding of the transpersonal, pointing out the need to distinctly differentiate the psychic from the magical and the subtle from the mythic. He makes a special plea that we take pains to recognize the wide disparity between transpersonal experiences and perceptions that are in fact pre-personal and often immature. He characterizes the latter behavior as being instinctual, infantile, impulsive, and self-assertive, traits more typical of regression to narcissistic absorption. However, the transpersonal, or transegoic, is an entirely different set of psychic structures that is characterized by an increasing self-realization and, in all, a more elevated spiritual state.

Charles Tart concludes that transpersonal experiences are potential in human beings and arise from the deep unconscious. His studies, he declares, strongly imply "that human consciousness may not always be restricted to the body and the brain," and make the case that "there may be other kinds of consciousness than human with which we can interact."[9]

Transpersonal consciousness, then, is a natural but latent human potential that is fully realized by only a few. Therapeutic Touch becomes an entry point as the therapist incorporates the TT process into her lifestyle and accesses deeper levels of consciousness.

Therapeutic Touch and Transpersonal Psychology

The transpersonal is always grounded by experience. For the therapist, this occurs during the enactment of the TT process. One can, in fact, see a progression in the stages of the transition to the transpersonal; however, the characteristic traits are not always acquired in a strictly logical succession.

First comes the recognition and implementation of the power of compassion as the TT therapist strives to help or heal those in need. Next, the purposeful use of breath during the act of intentionality serves to alter the biochemistry of the body. Intentionality is reinforced by the clear visualization that actualizes the healing moment.

Throughout, the continued centering of the TT therapist's consciousness in effect converts the Therapeutic Touch process into a mindful walking meditation. This act of interiority is attended by a sense of timelessness, at-one-ment with all beings, and a profound peace that seems enfolded in an atmosphere of utter stillness. In this milieu, direct contact with the healee's vital-energy field and, to an individually variable extent, the intermesh of psychodynamic fields, sensitizes the healer-healee interaction so that mind-to-mind communication can occur naturally and easily between both parties.

As the therapist progresses in depth in the practice of Therapeutic Touch, and her linkage with her inner self becomes clearer and more definitive, she achieves greater access to intuition in her attempts to help or to heal those in need. In fact, there seems to be a direct relationship between the degree to which one has access to intuition, and one's willingness to put these messages of personal knowledge into action in daily life.

As the TT therapist makes these attributes her own, and her experience presents other alternatives to her perceptions, her worldview changes dramatically to accommodate her glimpse of the multiple realities that have enlarged and vitalized her universe.

How Therapeutic Touch as a Lifestyle Forces Personal Transformation

Therapeutic Touch is, above all, a conscious, mindful process, which implies that a discriminating Self is involved in all facets of this healing interaction. Some features of this state of mindfulness can be perceived by astute observation and simple deduction.

The initiation of Therapeutic Touch itself is volitional, which suggests that judgment about the feasibility of success is in place. Intentionality, which is at the core of the rebalancing phase of Therapeutic Touch, is itself a practice of sharply focused awareness, and it takes a keenly aware person to fine-tune the level of visualization that serves to ground the intuition. Moreover, the reassessment of the healee's vital-energy field that concludes the TT session demands for each healee an incisive insight into the probable course of his or her healing.

A willingness to be changed or transformed by the experience marks the committed TT therapist, for a prerequisite to success in healing is a voluntary consent to go through many deep changes that may be evoked by the developing liaison with one's inner self. Indications of this deep touch appear naturally: there is a sharp increase in intuitive ability, a pronounced increase in confidence in TT skillfulness, and a significant change in worldview. These become evident in the therapist's acts of daily living, as well as during the TT interaction.

The personal transformation of the individual TT therapist is as diverse as the complexity of her personality and the intricacies of her life experiences. However, as the therapist comes to understand her intimate yet mysterious relationship with her inner self, there is a concomitant deepening sense of responsibility for relationships and a seemingly unaccountable increase in difficult choices in her life. One notes baffling behaviors in oneself: memories that skirt the edges of déjà vu arise to consciousness unexpectedly, or hitherto unnoticed relationships between concepts now promise to draw meaning from obscure situations, simultaneous happenings, or unexpected occurrences.

In fact, there are so many instances of the unexpected that they may seem to be a hint as to what is happening. Perhaps a reason for the sudden shift in events is that the conscious dimension of one's life has changed, and circumstances now have an opportunity to present themselves to a previously latent, newly quickened facet of one's consciousness.

Embarking on an inner journey to seek out the inner self as a ground of being is not for sissies! Actually, the belief that the inner life is accessible lends courage and decisive purpose to one's life. As a result of this liaison, the fruits of living become more meaningful, more interesting—and more fun.

Therefore, we can see that Therapeutic Touch is both a mode of healing and a support for the personal growth of the TT therapist. It is an opportunity for increased self-knowledge, increased apprecia-tion of subtle energy dynamics, and increased communication with the inner self. In the process of integrating the inner self, one becomes aware of multiple realities. Therefore, one needs to be well grounded, centered, cognizant of the inferences of one's actions, and accepting of that responsibility.

Therapeutic Touch becomes a transpersonal act as the therapist allows her inner life to permeate her outer life or persona. This rela-tionship with the inner self brings about a unique energetic, releas-ing the finer energies of the higher orders of self. We see this dynam-ic enacted most clearly through acts of compassion. As she identifies with the process, the TT therapist becomes increasingly sensitive to others. This constitutes a shift from a primary concern with self, and alters one's value system. When one draws upon the deeper levels of inner life, circumstances of one's outer life no longer relate to a sim-ple cause-effect standard. It is as though one enters a different time-space. The context of one's life changes. Priorities change. The mean-ing of everyday relationships changes. Probably for the first time since our Neanderthal ancestors, concern extends beyond one's personal family—and it is the power of compassion that provides the drive for this evolutionary leap.

As the healing way becomes one's lifeway, it is at the level of the inner self that both spirit and healing become personally meaningful. To a greater or lesser extent one begins to recognize indicators of the inner self in everyday life: there is a sense of equanimity and composure apparent in the therapist's demeanor, a stability and confidence marks her life, and she takes voluntary action to fulfill her obligations instantly and without rancor. She shows self-restraint, self-discipline, and a resolute persistence in meeting goals and fulfilling responsibilities. Most often she feels calm and unruffled; physically she acts as one piece, without clumsiness or hesitation. She maintains a sense of the future, knowing what she must do now in order to be ready at some future time.

The behaviors of the TT therapist and the healee during the TT interactive process itself have important consequences for building a bridge to the transpersonal. The TT process is usually so quiet and gentle that it is sometimes taken by the untrained observer to be a simple, mindless act. However, those taking part in the healing enactment often know it for the powerful multidimensional healing process that it is, and are able to help many people by adding creative personal touches to the TT process drawn from their experiential knowledge.

In this process there are at least three arenas for key events: within the TT therapist, within the healee, and in the interaction between them.

Within, the healer is motivated by compassion. This motivation bestows a kind of power that remains a mystery. It is exemplified by the ability to effortlessly attain states of consciousness that were outside of previous experience, and to function with mindful awareness of one's actions. Secondly, the healer is aligned in a centered state of consciousness, and it is in this state that one finds the power and commitment to forge a path to the inner self. The TT therapist uses intentionality in a knowledgeable fashion, which lends surety, stability, and confidence to the effective direction and modulation of the healing energies during the TT rebalancing phase. Her actions are oriented toward the higher orders of self, guiding her quest.

The therapist's orientation toward higher orders of self—structured by the sustained centering of consciousness that underlies the TT process—reflects upon her physically. In sustaining the centered state from the beginning to the end of the TT interaction, the therapist becomes aware of several attributes of that state: there is a psychomotor quieting of the physical body, a sense of pervading peace, an impression of timelessness, and an awareness of a profound stillness. As this centered state is maintained, there is a significant lessening of chatter in the brain. There is a shift in ego focus and a significant, often surprising increase in the simultaneity of life experiences. A change in worldview leads to a change in lifestyle. There is clarity in the recognition of compassion as power to help those in need. The healer is guided by a conscious mindfulness and a ready access to deep inquiry, and there is a tacit understanding of the effortless effort—the ability to do more than you ever expected to be able to do, with a minimum of exertion; the energy needed to accomplish physical tasks or spiritual quests seems to arise easily, as if from a bottomless well, as the higher chakras begin mature functioning.

During the TT engagement, the therapist has subjective reflections on past experiences, learned knowledge, built-in biases, the free association of ideas, and in-depth analysis of the ongoing TT process itself.

The healee, who seems to be just quietly sitting there, is actually the focus of another level of dynamism. The healee is often motivated by anxiety about the illness, which, if prolonged, will lead to increased stress. His mood stimulates his neuropeptides; endorphins and enkaphalins evoke responses in the immune system and repattern the subtle energies of the emotions, allowing the affect and mindset to shift. The therapist affects the nature of the healee's vital-energy flow, and resonant waves of endocrinal and other physiological changes can occur as the TT treatment progresses.

Meanwhile, a rapid relaxation response further reinforces the stimulation of immunologic response in the healee. There is a lessening of pain, which may seem to the healee like a spiritual experience. The healing process itself is accelerated, so that the healee has a sense of respite and can turn his or her attention to other things. These manifestations of the healing process may evoke in the healee a will to change the conditions of his life, or, when the lessons of the illness are learned, the healee may be free to move on.

These changes are significant, because they arise from a well-documented "sick pattern" of abnormal withdrawal to some inner place of self-consciousness, an exclusion or cutting-off of oneself from daily involvement in acts of living and being, so that the ill person is set apart from the naturally constant and potent interaction between the universe and the individual. In this way, the healee's linkage to the "whole" of life and living is abruptly truncated. This gives rise to a restlessness that may progress to a state of high anxiety, and a difficult-to-pin-down discomfort that may grow into a sense of intense pain. The healee may feel dogged by a gut-wrenching fear, a sense that things "just aren't right." He feels outside the loop and senses that he has lost control.

But as the healee recuperates, there is a dramatic change. His sense of well-being returns, and he feels in sync with the universe. There is a constant and consistent rhythm to daily life. He no longer needs to think about life's details; they just happen. There is a rhythm to daily life and the healee can "go with the flow," for events now have a predictability. As the healer projects or "reaches out" to the healee while anchored in the realm of the inner self, the healee can also become rooted in a timeless place. This sense of timelessness may seem to the healee to reflect help from "otherworldly" beings, an idea that may distract him from the reality of his condition; therefore one must use caution and intelligence in explaining the healee's condition to him.

In considering the interaction between healer and healee, one must take into consideration that both verbal and physical communication play a part. Jourard has documented that the healer in offering touch is saying, "I want to help you," and the healee in accepting touch is replying "I want to be helped." Where there is no physical contact, as, for instance, in the TT assessment, very often the healee will say something quite surprising, such as, "I've never been touched so deeply." And one begins to realize that body contact is not the only way we touch one another.

There is also an innate unitive experience that frequently occurs between the healer and the healee during the TT session. One therapist explained, "I feel a sense of stillness, peace, and an inner strength during this time of at-one-ment. I feel loving and accepting of both the healee and myself. I know I am helping her, and I am proud and pleased that I could do so; it feels so right, it makes me certain that this is what I was born to do."

The voices of healees who have been engaged in Therapeutic Touch echo these sentiments, and through their words it can be recognized that transformation to a transpersonal level can be a shared experience.

The following are direct quotes from healees:

- "The TT session filled me with a sense of gratitude and a reverential feeling difficult to describe."
- "I felt more closely in touch with the universal life force through this experience."
- "My point of view shifted; I was more optimistic about my life and my place in the universe."

- "I felt stronger, more stabilized within myself."
- "I felt more certain that I could overcome obstacles."
- "I had an increased sense of confidence."
- "I felt someone cared about me."
- "I felt an unanticipated sense of security."
- "I felt a weight lift from my shoulders."

These beautiful statements are profound. This is the voice of spirit in the midst of the healing moment, and the meaning is in the act itself: healing truly is the most humane of all human acts. It is a natural potential in all of us, and if one can do it, one should, for it is both a bridge to the transpersonal and a pathway to the self-realization of compassionate concern that helps us become "more fully human."

CHAPTER X: FIELDS IN WHICH WE LIVE: TOWARD A COMPREHENSION OF ENERGY FLOW

Human Energy

The concept of physical energy was introduced by G. W. Leibniz over three hundred years ago, and that idea was put to work on practical tasks by his contemporary, Isaac Newton. The prime characteristics of energy are that it flows or is continuous as it moves through space, that its flow has a coherence or rhythm, and that it has the capacity to do work. Its flow has been described on a continuum from slow to fast, strong to weak, unimpeded to congested, tenuous to thick, or quiet to tumultuous, depending upon the situation. Its rhythm has been characterized as steady or irregular, in harmony or unharmonious, in sync or disorganized. The work performed by energy can be calibrated, or measured.

Physical energy is essentially electrically charged particles that are expressions of force fields. It is these force fields that make up the fundamental building blocks of the universe, and their energetic flow can be converted into matter. At the matrix of this effort are information and consciousness. When human energies arise out of the physical body and endow it with the distinctive emotional, mental, psychic, and spiritual qualities that distinguish the individual organism as a human being, they act from the forces of consciousness. It is with these human energies that Therapeutic Touch achieves its healing effects. One might say that healing, along with a few other primal and creative forces, such as music and mathematics, is a humanization of energy. From a human energy perspective, illness is an energy imbalance; weakness is a significant energy deficit or energy loss; and disease usually is due to an energy blockage that results in uncontrolled hyperactivity or repression and the loss of synchrony in the vital-energy flow.

Recall that field theory is the central model of the TT conceptual framework. The human energy field as it relates to the individual could be called a localized field, and its various levels of consciousness termed subsets of that field. These subsets are the vital-energy level, the psychodynamic level, the conceptual level, and what might be called a spiritual level of consciousness. Because these levels of consciousness have multiple facets that seem to act in a unified way, they too have been called "fields."

The innate characteristics of the human energy field that allow the healing process are its self-organizing capacity, its ability to replenish energy (during sleep or wound healing, for example), and the interpenetration of energy levels. These crucial elements in turn are grounded by prana, or vital-energy. Embedded in the physical matter of the individual, a restless, rhythmic, sensitive, and reactive vital energy flows at the structural core of each cell. As these cells divide and multiply into tissue, and as the tissues form into organs, these vital-energy flows translate into fundamental frequencies that act as a vibratory signature for the individual. When these oscillations are concordant or in balance, there is health; when they are incoherent or out of phase, there is illness. Swami Rama, a world-renowned teacher of yoga, suggests that the hairlines of the body are patterned along the symmetrical paths of vital-energy flow.[1]

Interpenetrating each person's physical matter and vital-energy flows is his or her psychodynamic field, the bearer of such traits as sensory perceptions, reason, and the emotions. Kunz states that the psychodynamic field is more permeable than the vital-energy field and can exhibit great elasticity or pliancy, so that powerful discharges of feelings or thoughts dramatically enlarge the field. Pronounced flexibility and malleability is a distinctive feature of healers' psychodynamic fields. Kunz reports that a strongly emotional cascading of psychodynamic energies can interpenetrate the fields of other people in the vicinity at that time, producing a charismatic effect or other strong influence.[2]

Both the vital-energy and the psychodynamic fields are further interpenetrated by the individual's conceptual field. It is characterized by its intellectual functioning: the ability to conceptualize, synthesize, and have intuitive insights and clear visualizations. In a word, the conceptual field distills meaning from one's experiences, and as the meaning is illuminated by intuitive insight, one has the opportunity to grasp the creative moment.

Finally, the spiritual field offers an "open sesame" to the sacred and the metaphysical. As the domain of the sublime, it offers experiences that cannot be conveyed in words without the nod of the Muse.

As previously noted, all of these localized fields are finely integrated and elegantly orchestrated so that the individual can perform her daily activities as a unified whole.

Human Energy as Pattern

Healing is a process that draws upon the dynamic circulation of prana to form harmonious patterns involving the body and, more importantly, the mind. The renewal and invigoration of this harmonious circulation of prana occurs in many methods of healing, such as song, dance, drumming, and the entering of a mandala, painting, or medicine wheel. In Therapeutic Touch it occurs through touch, both tactile and nontactile.

The TT assessment process is in fact a direct experience of another's vital-energy field and, for a sensitive few, the psychodynamic field as well. It is during the assessment phase that the healee's energetic flows are translated in the mind, via the hand chakras, into a language the brain understands. This does not mean that the translation is in English; in fact, English is woefully deficient in descriptors of the nonphysical. As discussed above, a glossary of metaphors has emerged among TT therapists to describe these nonphysical impressions that act as cues to the state of energy balance in the healee's fields. Again, touch "speaks" to the TT therapist by direct perceptions or impressions of energy flow patterns as:

- Differences in temperature
- A sense of attraction, similar to magnetic pull, or repulsion
- Cognizance of energy deficits or of hyperactivity
- Congestion or blockages of energy flow
- Sensation of tingling or weak electric shocks
- Pulsations or an awareness of dissonant rhythms
- Intuitive insights
- Direct communications from one's own chakra complex

There are also distinguishing qualitative aspects to the cues that signal emotional connotations of the energy flows. These have been expressed as

vibrancy	depression	"nobody at home"
overflowing	depletion	fear
joie de vivre	emptiness	panic
restlessness	confusion	anxiety

Some TT therapists also use personal codifications to clarify or explain their individual comprehension of these flow patterns to others or, frequently, to themselves. They usually do this by voicing sounds or by adapting symbols, private language, or other expressions in an attempt to bring to common understanding their experiences with these subtle energies, which are not well understood in our time.

Modifiers of Human Energy

As Therapeutic Touch becomes part of one's lifestyle, the sensitive therapist becomes aware that there are many little-recognized factors that can modify one's own energy field. High on the list are people themselves—so-called sappers and slurpers—who by their nature drain off the vital energy of those nearby. There are also are high-energy people, the uplifters and the stimulators, whose altruism, charisma, or joy serves to boost others' energy levels and raise their spirits.

Animals are strong modifiers of vital energy—particularly cats, who were considered to be of divine origin by the ancient Egyptians by virtue of their strong pranic outflows. Dogs and horses willingly share their emotional energies, particularly love and adoration. Some birds (especially hummingbirds, hawks, and eagles) and sea animals, such as dolphins and whales, can act as catalysts for human emotional and conceptual transformations. Looking into the eye of a dolphin or whale as one swims alongside it is like looking into the soul of the universe, and one is mightily changed by the experience.

However, certain geographical sites—called sacred places—are the most powerful modifiers of vital-energy flows. Sacred places are noted for their invigorating and therapeutic effects; for example, high in the mountains, at sites of running water or heavy tree growth, or at the seashore. The common physical factor that has been noted at all such sacred places is a high concentration of negative ions, which can act as carriers of prana, in their atmosphere.

Such nonliving factors as music or color, and even abstract elements such as direction (e.g., feng shui), can also modify vital energy.

Effects of Emotions on the Psychodynamic Field

While there is no objectivity in psychic space, it is possible to find constant patterns in the psychodynamic field. These behavioral patterns are known as emotions, emotional dispositions, or moods. Stress, of course, is one of the most potent modifiers of human energy, and can easily transform psychodynamic energies into negative excesses such as anger, rage, anxiety, apprehension, and fear.

It is said that stress is pandemic throughout the world and causes psychosomatic illness, which accounts for seventy to ninety percent of all illnesses. All of the stress-induced emotions noted above respond very well to Therapeutic Touch, usually as a result of the very strong relaxation response that occurs within two to four minutes of starting the TT process.

Whether positive and elevating or negative and oppressive, forceful emotions are "contagious" to others in the vicinity. It is important, therefore, to break the mold in which adverse emotional patterns are shaped and to nullify persistent patterns of automatic unconscious reactions. In the United States at the time of this writing, automatic emotional reactions are epidemic. Resentment, which often occurs when anger is not acknowledged, is now the dominant adverse emotional characteristic of North Americans. One can easily visualize the insidious creep of those dangerous emotions: anger—resentment—road rage . . . remorse.

The persistence of these progressing negative emotional patterns cannot be overstressed. We know from Dora Kunz's studies of healees during Therapeutic Touch sessions that patterns of psychodynamic energies, such as depression, fear, and tension, all tend to diminish the normal flow of vital energies. In cases where hysteria or panic are superimposed, the person's vital-energy reserves can be depleted. When this happens suddenly, the patterning can shift too rapidly, and heart attack or kidney failure may result. Or a "spasm" may occur in the energy flow to the liver, producing a shock to the healee's vital systems. It also has been found that interference with the rhythm of vital-energy flow is immediately felt in the heart, the respiratory organs, or the gastrointestinal tract, sometimes with subsequent pathological sequelae.

These dismal emotional energies are destructive in many ways. Patterned as anger, they drain vital energy and disturb intrinsic rhythmic patterns; constellated as fear, they immobilize; as pain, they congest and stall the vital-energy flow; repeatedly mulled over as resentment, they slowly erode the essential flow and rhythmicity of vital-energies; guilt stifles spontaneity and creativity; and long-standing and excessive grief can dissociate or actually disconnect the bereaved person from the vital flow, so that physiological and psychological illness may occur.[3]

Dora Kunz has offered several cogent suggestions for those TT therapists who are trying to restructure their lifestyle to better help or heal persons in need. Since a therapist must maintain a wholesome flexibility, adaptability, and responsiveness to her own vital-energy field, she may find of particular interest those suggestions concerning emotional behaviors that become fixed and firmly rooted in routine and repetitive acts. Kunz has pointed out that when feelings are habitual, they become embedded in the energetic structure of our being. Over time they become intrinsic to our subtle energy fields and, as a built-in arrangement in the field, are related to everything we do. As the individual habituates to that emotional pattern, these feelings seem to him to be deep-seated hereditary traits, an essential part of him. Once one recognizes the habit, Kunz recommends centering and becoming consciously aware of the unintentional reflex feeling or action. This shift of consciousness temporarily stops the automaticity of the patterned emotional response, providing an opportunity to begin a self-awakening in which we can call upon the processes of self-healing.

From the Eastern point of view, there are three major types of energy concerned with health: prana or vital energy, which is concerned with the energy underlying the organization of the life process; kundalini, which is involved with a broad spectrum, from creative energies to the libido; and a third category, called by its Greek name, *eros*. The TT therapist accesses kundalini over time, particularly when she learns to use her chakras in a responsible and therapeutic way. Eros is concerned with love in its many guises, and is one of the most powerful motivating forces of the therapist. In a long-term study on Therapeutic Touch reported elsewhere,[4] it was found that the most effective energies projected by the TT therapist to dispel imbalance is the power of love.

Prana, or vital energy, constantly flows through us to keep us lively and healthy, and is the foundational energy used in Therapeutic Touch. Vital energy is the prime life energy. Its distinguishing traits are vitality and vigor. It is the fundamental vibrant source of animation, the dynamism and élan that mark the living and generate physiological functions such as movement, rhythm, and the growth and development process. Of these functions, rhythm is the most essential characteristic of vital energy. It turns up as peristalsis, respiratory cycles, muscular movement, and brain waves—and Therapeutic Touch is very effective in all four of these areas. It is highly successful in the healing of paralytic ileus, in which there is a cessation of peristalsis following abdominal surgery; in the alleviation and recovery of upper respiratory illnesses; in relieving or eliminating muscular pain; and in evoking the relaxation response, as can be seen by the alpha brain waves.

We achieve this not so much with the amount or intensity of the vital-energy projection we direct toward the healee as by integrating the healee's energies into a synchronous flow. In Therapeutic Touch, our objective is to work with the healee toward vital-energy balance, where all significant subsystems are finely integrated into one harmonic whole.

CHAPTER XI: ENERGETIC FORCES AT WORK DURING THE THERAPEUTIC TOUCH PROCESS

We now realize that the universe is inconceivably energy-rich, and that this high-powered milieu is in constant interchange with everything that incorporates matter into its being—the animate and the inanimate, the swift and the halt—and that each type of energy in this interactive convergence is regulated by its own dynamic self-organization.

Within this constantly moving universe, I have identified five forces that significantly affect the TT process (recognizing that there may be others). For the purposes of this discussion, these five forces will be assigned the letters A through D and X, in the following manner:

Force A will be assumed to be generated by the higher orders of the TT therapist, the inner self that is committed to helping those in need.

Force B will be relative to the state of consciousness of the healee and the therapeutic interaction with the TT therapist.

Force C will be generated by the circumstances in which the TT session takes place; e.g., sacred places that are oxygen-rich and therefore have abundant supplies of prana.

Force D will be produced by the direct intervention of natural powers, of which we are aware, but about which we understand little and publicly admit to less. I refer here to the angels or spirit guides that are mentioned in the sacred books of all the major religions and many of the minor ones, and to certain trees whose consciousness can be solicited for help if they are in the vicinity of the TT process as it takes place. Those of us who lack the ability to directly perceive such intercession can verify the success of their intentionality by recording the physical effect on the healee.

Force X, the group for miscellaneous items and non-local effects, can be concerned with factors such as the thoughts of loved relatives and friends, healing groups, or other people, and additional subtle energy factors of which we are unaware.

Even with only five field forces, the healing milieu surrounding the TT session is quite complex. Simple graphic depiction serves to clarify it somewhat (see Figure 1).

Figure 1

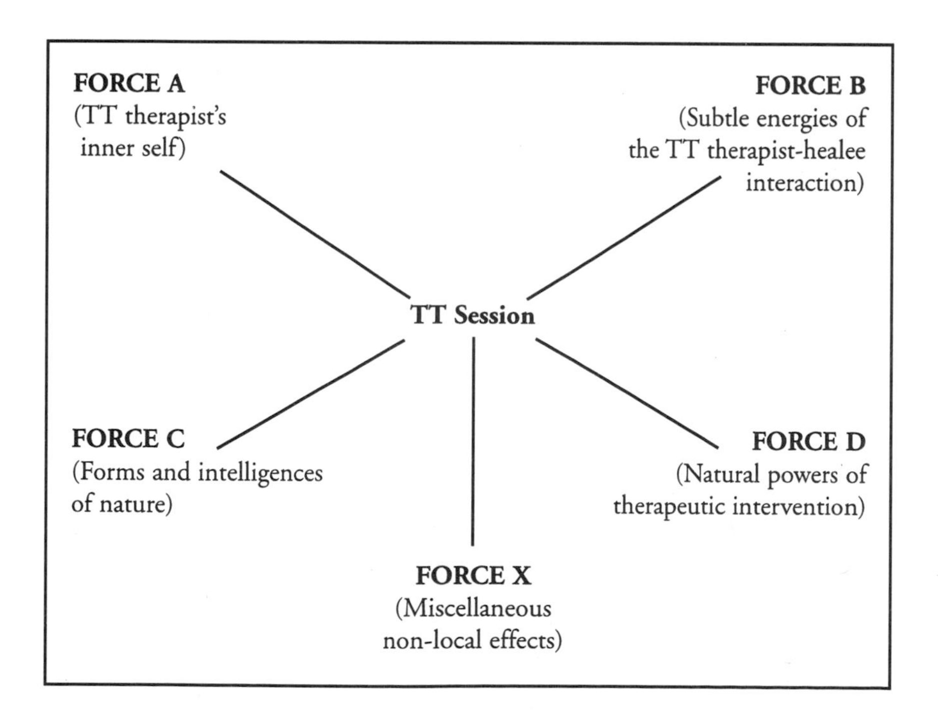

Force A: The TT Therapist's Inner Self

Centering, the "Open Sesame!"

It is the centering of the consciousness that provides the stillpoint on which the inner self can effortlessly enact its unique talents. To achieve this, we must realize that most of us are incessantly interacting with others and with various aspects of ourselves: in this competitive world our vital energies are constantly engaged in interactive one-upmanship encounters with those we'd like to impress, or with ourselves. However, once we engage in an attempt to help or heal someone in need, we have to relinquish this often routine exchange and quietly go within. From that stillpoint, the TT therapist feels the serenity of her unified being and can project that sense of deep peace to the healee.

This entails a non-ordinary kind of intelligence and calls upon non-typical ways of communicating that involve the still poorly understood functions of the psyche. Centering in TT becomes an arena for the self-exploration of the therapist's human energy fields, which she undertakes without anticipation or expectation, allowing the experience to arise out of its own being.

How does the TT therapist achieve a centered state of consciousness? Motivated by compassion to help someone in need, she quiets herself and turns her attention within. The energy dynamic is opposite to empathy, which involves a flow of the vital-energies out to the healee; in centering, the healer brings her energies inward to the stillpoint. At first, she feels the everyday pulsations of life in her environment and becomes aware of her reactions to them. At some personally determined point, which is usually within a very few minutes of acknowledging her intentionality to center, the TT therapist experiences a felt shift in her consciousness that she recognizes as a sign of "presence"—that her total self is at attention, that it is "there." She focuses on that new state of consciousness, and in a short time has a sense of nonverbal communication. She directs her attention inward and "listens" intently for a response, often getting caught up in this different reality. At this time the TT therapist becomes more aware of the natural, invisible forces around her.

As the content of this inner communion becomes clearer, she deepens the level of her attention, following the dominant thrust or reach of the response from within toward a higher or deeper level of consciousness. She rapidly and sensitively moves her hand chakras through the healee's vital-energy field in a TT assessment, using her centered state as a gauge for all that is not in balance. As the information she picks up during this assessment becomes clearer, she sharpens the focus of her attention, still following the leading edge of the response from within and preparing to use this singular energy to help the healee. She observes her own inner dynamics as objectively as a spectator who has come to learn, by deliberately shifting her focus from what is going on around her and concentrating on her subjective senses.

In the course of this compassionate work, the therapist experiences heightened feelings of harmony, balance, and integration of the senses, producing a clearer perception of the dynamics of the therapeutic interaction in which she and the healee are engaged. If she is able to maintain that state of centered consciousness, she finds that the deeper she can journey within, staying on center and anchoring her consciousness as she heals, the easier it is to help those in need.

What does the centering process teach the TT therapist about her inner self? In centering, the therapist attends to a "feeling" of deep inner quietude; she experiences this state as a reality and focuses on the sensations emanating from it. She follows those sensations inward "as a thread"; that is, she identifies with her inner quietude and expresses or projects it through the TT process to meet the healee's needs. At one with the mind, the therapist's hand chakras seek the sites of imbalance. Nature literally talks to us in patterns, and the therapist may get a strong sense of what these patterns of cues, the flows of vital-energies, are saying. The translation depends on the nature of mind that has been developed and the nature of one's linkage with the inner self. For instance, there seems to be a feedback loop with the therapist's state of confidence that affects the clarity of her interpretation of her communications with her inner self. Her certainty and clarity seem to be sharpened by personal meditative practices and, in some inexplicable way, the interchange with her inner self seems to be supported and sustained by the depth of her compassion for the healee. The communication or "translation" often occurs with the immediacy of intuition, providing information about the state of the healee's vital-energy field and somehow structuring the TT therapist's act of intentionality in reference to how the healee's vital-energy field should be rebalanced. Indeed, it seems that the healing itself is embedded in the in-flowing process.

Centering, therefore, has dual functions in the Therapeutic Touch process. It provides entrée to the TT therapist-healee interaction, and supports that interaction by sustaining an appropriate milieu for the direct participation of the TT therapist's inner self. The preparatory factors that the therapist brings to Force A's effect on the healing moment, as indicated in Figure 1, are characterized by:

- Psychomotor quieting of the physical body
- A significant lessening of uncontrollable verbalizations ("monkey chatter") in the brain
- A sense of timelessness
- A profound stillness and sense of peace
- A diminution or shift of egocentric focus
- Clarity in the recognition of compassion as power
- A deepening appreciation of conscious mindfulness
- An increase in self-confidence
- A strong grasp of intuitive insights

- Access to deep inquiry
- A tacit understanding of the power of effortless effort; e.g., when working from the bramachandra chakra, at the back of the top of the head.

Clothed in these attributes, the centered TT therapist has little difficulty in staying on-point, finely focused, and able to force coherence out of the constant flow of memories, ideas, metaphors, information, and context that usually clutter the mind. It is because she has prepared that she is able to provide the conditions of alignment and quietude in which the personality's liaison with the inner self flourishes.

Centering has a profound but individually varied effect. The following verbatim descriptions are presented in an attempt to capture some of the TT therapist's interpretation of this personal experience:

- "I felt I was in a different place than my usual physical surroundings, a special place that I almost remember."
- "I was able to 'tune out' activity in the environment; I was aware only of the utter stillness of the moment."
- "It was as if I was alone with the healee and I had a deep, true knowledge of what had to be done."
- "I felt I had been 'there' before; it was a very familiar feeling, as though I was in touch with a wise friend."
- "It was as though a sense of quietude and peace pervaded my being, which I felt emotionally as well as physically."
- "In that moment, my love for the healee was so profound, there was no question of whether I would succeed in helping him. I knew that at the least, I had to try."

The Experience of the Inner Self

There are certain indicators of the inner self that have endured through time. Early on, one becomes aware of a substantial increase in personal knowledge, an interior knowing that can be trusted. This awareness is introduced by a felt shift in consciousness that serves to ratchet up the awareness of nonphysical cues. Several things seem to happen at once: there is a calming and quieting of the physical body, and a unique stillness seems to permeate the atmosphere. One feels a sense of peace, of equilibrium and well-being. There is also a sense of timelessness, as if time is no longer an appropriate consideration. One's perception deepens, and the nonlinear—waves, spirals, Möbius happenings, and the timeless—rule, instead of stark rectangularity and tick-tock time. Synchronicity enters one's life and power appears, but on its own terms. There is an irrepressible welling up of desire to elevate one's level of behavior, to surmount disabilities and negative patterns of behavior, overcome crises, and realize a sense of direction in one's life. Thus the gifts of the inner self resemble those of centering but extend past the TT process into daily life.

The Inner Self As Ally

It is strange that our current culture, the most technologically advanced society known to this era, has been marked by a deep remembering of healings that have their roots in the ancient past. And the outstanding feature of these ancient healings is the deep appreciation that it is the inner self, the true spirit, that is the actual healer. Recruiting the inner self's active participation in one's life therefore becomes a crucial goal for the TT therapist.

The way is well marked for those who wish to gain access to the inner self; centering the consciousness is the key. The earnest pursuit of this goal cultivates a personal sense of peace and equanimity, a disciplined sense of inquiry, and a steadfast determination to bring forth this elegant, insightful, and benevolent aspect of one's being.

The process entails an individual commitment to follow through on the prompting of intuition, coupled with a willingness to recognize and acknowledge the differences between true intuition, hunch, and self-will, and to act always with the highest resolve.

Once one starts to practice Therapeutic Touch with serious intent, it becomes apparent that Therapeutic Touch as a lifestyle forces transformation. One can watch this radical change happen to the TT therapist before one's eyes, so to speak; it is an incredibly rapid and dramatic remodeling of the self from the as-usual daily agenda of most of us, as can be seen from Table II, below.

Table II. Attributes of the Inner Self vs. the Social Mores of Our Time

SOCIAL MORES OF OUR TIME	LIFESTYLE OF THE COMMITTED TT THERAPIST
Competition	Mindful helping/healing
Complicity	At-one-ment
Aggression	Peace and serenity
Self-aggrandizement	Unitive nature of all living beings
Restlessness	Stillness, inner quietude

Clearly, the TT therapist aspires toward spirituality. This quest is in the mainstream of our time; a 1997 Gallup Poll reports that eighty-two percent of those surveyed said that spiritual growth was a very important part of their lives, and a quest for the spiritual becomes more evident every day in some critical avenue of American culture. This indicates that a quest for the inner self (a native of the realm of the spirit) in our time could be considered an inspired goal. From the spiritual point of view, all things are connected (people, other animate beings, and even beings formerly considered inanimate, such as mountains, rivers, rocks, and other offspring of Mother Gaia), death is a part of the continuum of the life process, compassion is a natural human attribute, and selfless service is important. All of these considerations can be strengths for the committed TT therapist.

However, the nearest idea that most of us have of the inner self occurs during those rare moments when we are gifted with inner promptings about some affair in which we are interested. Happily, this subtle realm is the playground of the TT experience; it is the compassionate urge to help someone in need that fires the merging with the inner self.

We are now at a moment in history when it is essential that we be responsive to the inner self, give it voice, and admit that it exists and can involve itself here and now. The inner self accompanies one into everyday affairs. One no longer feels afraid; one finds oneself taking the path less often trod. Mindfulness, considered observation, and insightful deductions come to mind with remarkable ease. In critical times, one finds an inner source of confidence, a sense of equanimity and composure, a quiet joyfulness, and often a presence of mind that startles and delights. There is an acceptance of the present and a sense of the future that helps one to act as one piece, calm and unruffled by exceptional events and able to face tomorrow with considerable restraint and an unflappable composure.

It is well worth the effort of grounding the inner self in one's life; in fact, one may find it useful to design simple tests for evidence of the effects of the inner self on acts of daily living. This can be done by carefully examining the experiential world for significant changes in relationships and other deeply meaningful interactions, dialogue, creative expressions, and involvement in community with others. Specifically, one could test for major changes in one's use of concepts, choices, goals, values, willingness to accept responsibility, and respect for the unusual individual. Such self-tests could also help other seekers to bring meaning and equipoise into their lives. Through another's example, they would learn to recognize, gain conscious access to, and nurture their alliance with the inner self—a process that would color the atmosphere of the healing moment, helping the healee as well as the healer.

Intentionality as a Focal Point for the Inner Self

In TT practice, intentionality is based on two factors: a specified goal, which arises out of and depends on the clarity of the TT therapist's assessment of the healee's vital-energy field, and an intense desire to achieve that goal. It is the basis for the rebalancing phase of Therapeutic Touch, the actual enactment of the healing process. Intentionality implies not only a wish to heal, but a strength of purpose, a strong commitment to help the person in need, and the ability to persist to that end. Related to free will and a formidable sense of self, intentionality also affects what is perceived and how one interprets one's experience. In Therapeutic Touch, it is the therapist's intentionality that impresses and directs the repatternings of the healee's vital-energy field during treatment, to bring the healee into balance.

Intentionality as a process is powered by a compassionate yearning to help those in need, and it is this strong desire that carries the consciousness of the TT therapist ever deeper. Physically, it involves a conscious use of the breath by the TT therapist on the exhalation. The "kinesiology" of this subtle energy response is action at a distance, similar to sending a basketball that hovers on the rim into the basket. This feat occurs through the instinctual use of "body English," a primitive, potent contortion of the body that expressively projects an intent or forceful wish; however, in Therapeutic Touch, intentionality requires neither stress nor strain. Instead of using the furrows of one's brow, the TT therapist uses visualizations of what she is trying to accomplish, reflecting these visualizations without strain onto the back part of the brain, whose functions are influenced by the brahmarandhra chakra. In this manner, the act of expressing intentionality overrides the mental "chatter" that uselessly fritters away vital energy, and allows the focusing of the healing process.

Intentionality has several explicit uses during the rebalancing phase of Therapeutic Touch. Primarily, it serves to align the TT therapist's consciousness with her inner self so that she can be fully present to the healee during the TT process. While in this mode it is the intentionality of the focused TT therapist that facilitates or stimulates the vital energy of the healee, mobilizing the sense of stagnation, congestion, or pressure the therapist has picked up in the healee's vital-energy field. Other functions of intentionality include the modulating, dampening, or quieting of the vital-energy flow, or the synchronization of its rhythmicity. Intentionality is also used in mind-to-mind communication between the TT therapist and the healee.

Vivid Visualization as an Expression of Intentionality

As stated in Chapter II, visualization differs from imagery. Imagery is an invoked mental image or representation of some imaginary object that an individual is able to create in his mind, such as the Wise Old Man, the Guide, or the Crone—an imaginary mental picture. The person practicing visualization, however, actually perceives the object in question, if in a manner more related to intuition than to vision.

True visualization may occur whether the object of it is in the vicinity or not. In Therapeutic Touch, visualization often happens during the assessment phase, while the therapist is exploring the vital-energy field of the healee without making direct contact with his body. As she uses her hand chakras to search the healee's subtle fields for indications of vital-energy imbalance, the therapist may receive a barrage of data at once. Because she is so highly sensitized, the total sensory input may constellate of its own accord into a clear visualization of the object of her search, providing her with a lucid perception of the circumstances under consideration.

Neither space nor time seems to be of much significance under these conditions; the entire event happens in the therapist's consciousness, and, like intuition, it carries with it a sense of indubitable assurance and authority. Indeed, the visualization invariably turns out to be true. This ability to correctly perceive at a distance has excited much interest. The first to formally study this phenomenon in our time were Puthoff and Targ.[1] I wondered if their interesting study could be applied to healing at a distance and subsequently conducted controlled studies, using professional nurses and their hospitalized patients. I called the process Vivid Visualization and reported on the results.[2] I will mention these studies here only briefly and refer you to the references cited for a fuller review of the theoretical rationale, presentation, and analysis of the research data, recommendations for further research, etc.

Vivid Visualization and Therapeutic Touch

Professional nursing care can be a highly personalized interaction between nurse and patient, with a consequent development of strong feelings of concern about the patient's well-being. Sometimes this concern enters the nurse's personal life and may even bridge the boundaries between her conscious and unconscious mind to enter her dream life. This disquietude may then rise to consciousness while the nurse is engaged in routine tasks or experiencing other frames of mind that are permissive of such perceptions. On such occasions, the nurse and the patient can be in direct mind-to-mind communication, and the nurse often can sense the welfare and activities of the patient.

I myself have had this experience. Three of these incidents gave me information about critical emergency conditions in which a patient was involved. Acting on this information, even though these patients were a considerable distance away, I telephoned the supervisors on call in two cases, and in the third case actually jumped into the car after midnight, then ran through the darkened hospital corridors to the patient's bedside in time to avert a very serious emergency. Episodes such as these gave me a sense of the validity of my visualizations, and I wondered whether other nurses had had similar experiences. Acting on this hunch, I did an initial inquiry in a random selection of 1,500 professional nurses in various parts of the country.

In this pilot study, about thirty percent of the sample (460) reported that they had had strong visualizations of their patients when they were spatially distant from one another. Of this number, eighty-two percent (377) stated that they had also had the experience of unexpectedly perceiving a visual image "within the mind's eye" of a sudden, crucial change in a patient's health status. Although they were unable to account for these perceptions in a conventional, rational manner, at a later date they found that the events they had visualized actually had occurred. At the time, these nurses had no professionally acceptable way of communicating this information to their peers, since there was no valid nursing theory that accounted for information transfers of this kind, and therefore most of them said nothing about these incidents.

Recent studies had indicated a significantly high correlation between compassion and psi, an information-gathering process that occurs via no known sensory receptors.[3] It seemed plausible that nurses working under the impetus of compassionate concern for people who were ill might have a high capacity for psi, and I decided to do a full-scale study on the factor, which I call Vivid Visualization.

In its final form, the study had experimental and control groups of professional nurses, and it was designed to test the reliability of their experiences with Vivid Visualization. The specific goals of this study were twofold:

1. To test the reliability of nurses' Vivid Visualizations about hospitalized patients upon whom they visualized themselves practicing Therapeutic Touch from a place geographically removed from the patients, and
2. To evaluate this talent for Vivid Visualization against the ability of other nurses of similar experiential and educational background who would imagine the conditions, surroundings, and interactions of comparable patients hospitalized in a remote location.

The Experimental Group consisted of fifteen teams. Each team was made up of a Nurse-Meditator, who did healing at a distance, visualizing herself doing Therapeutic Touch successively at the bedside of each of two patients, and a Nurse-Observer, who directly observed the patients during the Nurse-Meditator's visualizations. The Control Group was similarly designed, except that the counterpart of the Experimental Nurse-Meditators imagined the patients, rather than doing a meditative TT healing at a distance as described above, while Control Group Nurse-Observers monitored the patients.

All patients signed informed consent forms, and the study was approved by the various hospitals' research groups. In both cases the names and medical diagnoses were known, and the individual nurses in both groups were given specifically designed written protocols to follow. The study was done on three consecutive days at a mutually agreed-upon time. The entire procedure took twenty to thirty minutes each day. During the time of the study there was no communication between the nurses on each team.

At the end of the three days, the nurses on each team met with me. We reviewed their written reports, and the consequent discussions were tape recorded. These were transcribed later and rated, as were the written reports. These ratings were then objectively evaluated by a panel of experts who otherwise were not involved in the study.

The level of significance for testing the hypotheses was set at $P < .05$, that is, the probability of the results happening by chance was less than five times in one hundred occurrences. All the hypotheses actually exceeded that level; therefore, there was a high confidence that the nurses in the Experimental Group doing Vivid Visualization were correct in their visualizations significantly more frequently than the Control Group, and that the reliability of their reports was decidedly more dependable.

This study has been replicated several times, using other nurses and other health-care professionals, and the results were comparable. These controlled replications not only speak to the reliability of the study, but also force the recognition of how little we understand about the capabilities of human consciousness. The significant questions we each must face become clear: Are we willing to accept such power as a natural human potential? And, perhaps more importantly: Are we ready for the therapeutic use of the "paranormal," the psychic handmaiden of the inner self?

This is a time of great opportunity. We cannot allow our nerve to fail in acknowledging the therapeutic capacity of the exceptional reaches of the full human stature, a goal toward which the forces of Gaia urgently press. The inner self is our prime ally in the travail of human evolution. If we are truly moved by compassion to help or to heal those in need, then we must permit the inner self to participate in that healing moment. The crucial message is this: It is the expression of the inner self that creates the ambience in which each of us, as healer, can be totally present to the healee. That is the mystery and the magic, and the essential you, the inner self, is the magical child.

Force B: The Subtle Energies of the Healer-Healee Interaction

States of Consciousness during Therapeutic Touch

We can see that once we have given permission for the inner self to participate in the healing interaction, the practice of TT offers a new and different consciousness. For instance, when the TT therapist explores the healee's vital-energy field without making body contact, picking up cues about the healee's condition without having to ask the healee, the use of her chakras in this unique manner is a new way of social communication, a kind of intelligence that is atypical for most people.

Having created, through the act of centering, the appropriate milieu for the involvement of the inner self in the healing process, the therapist attempts to draw upon profound levels of consciousness as the healing proceeds. Dora Kunz has remarked that vital energy at the deeper or finer levels has substantial powers of organization. If the TT therapist is working at that unitive level, then she is working on all levels of consciousness at once. This gives her powers of assessment an added boost, as with practice she may be able to become aware of more than one field at the same time. For instance, she may become cognizant of the emotional and mental correlates of an illness while she is forming ideas about the physical aspects of that disorder via assessment of the vital-energy field. This power of sustained centering also begins to sharpen the TT therapist's intuitive abilities and her insight into problems. Therefore, if the TT therapist truly centers her consciousness even for a few moments, she may have an enormous effect on the healee.

Experience at these deeper levels of consciousness is very difficult to describe in conventional terms. This remarkable gateway to exceptional experience is well known by some, suspected by a few, and accepted by fewer, for this uncommon energy source is largely ignored in our culture; but you, the reader, must study it diligently if you would understand Therapeutic Touch.

Photos of the Sequential Process of Therapeutic Touch

To get a sense of the transpersonal quality of the TT interaction in the hands of experienced TT therapists, follow the sequence of candid photos of TT therapists during a TT session, on pages 128–138. To sharpen your personal identification with the TT interactions shown, take a moment to center before looking at the photos, and try to recapture the experience of the last time you did TT by closing your eyes and visualizing it. Try to feel as though you were there. When you are able to summon the events to mind and have a clear sense of that experience, examine the series of photos and the accompanying text, and compare your experience with that of the TT therapists in the photos.

1. Two advanced TT therapists, Jan and Stacey, reach out with their hand chakras toward one another as they center, preparatory to doing Therapeutic Touch. Notice the initial tension in the healee's body.

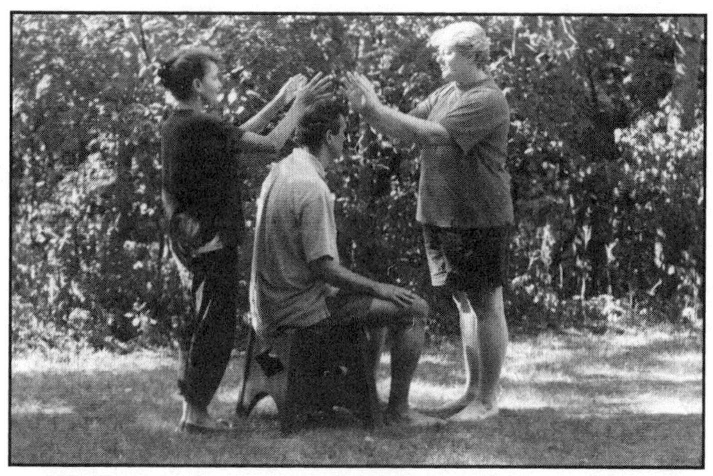

2. Staying in a centered state of consciousness, they begin to do an assessment of the healee's vital-energy fields. Note that their hands are not touching the healee's body. Although the TT therapists may make body contact as needed, one often becomes more sensitive when working only in the vital-energy field and not making body contact. Their gaze is centered, and their attention focused, on the hand chakras.

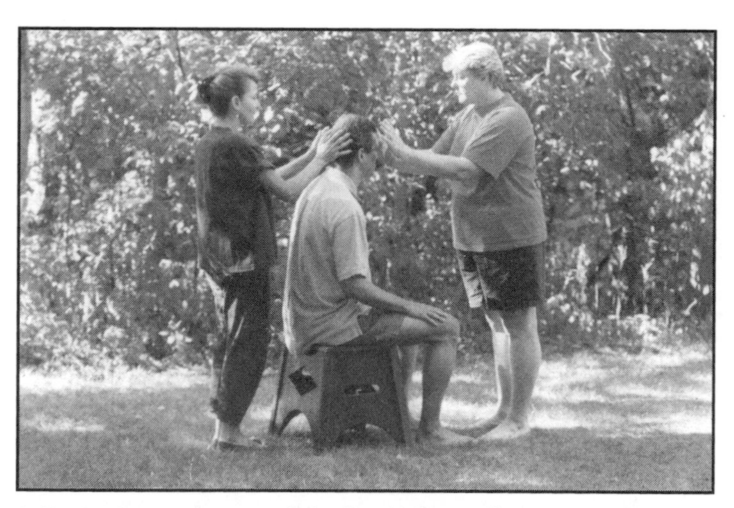

3. Beginning at the top of the head, they will systematically search the field downward, toward the feet. As they pick up cues from the healee's vital-energy field, they will store that data in their minds until the conclusion of their individual assessments, when they will share information with each other.

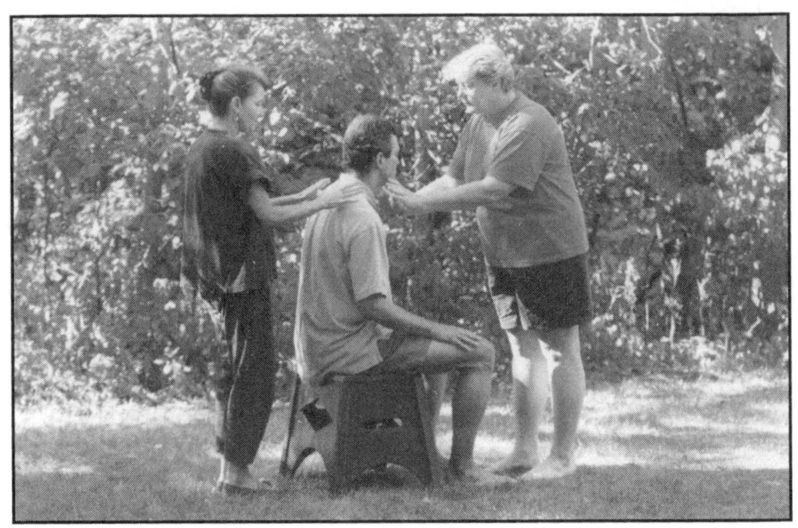

4. The therapists' concentrated gaze indicates that they are engrossed in deciphering the cues they are picking up from the healee's vital-energy field.

5. The TT therapists assume comfortable body positions as they continue to assess the lower part of the healee's vital-energy field.

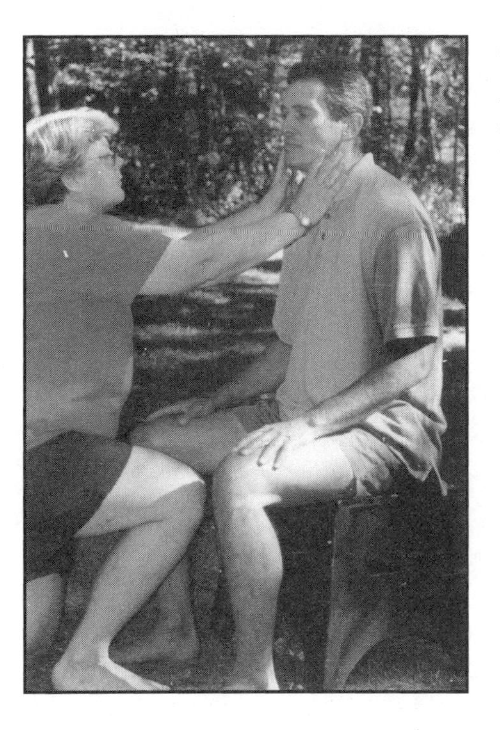

6. Stacey assesses the healee's lymphatic system in the neck, acutely aware of any difference she feels between the two sides.

7. Stacey assesses the healee's vital-energy field overlying the chest area, where she will be aware of any deviations from bilateral symmetry of temperature, energy flow, pulsation or rhythm, or visualization.

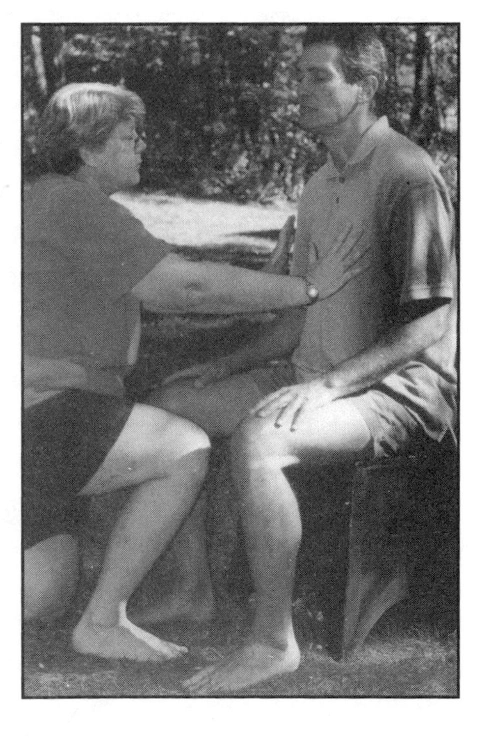

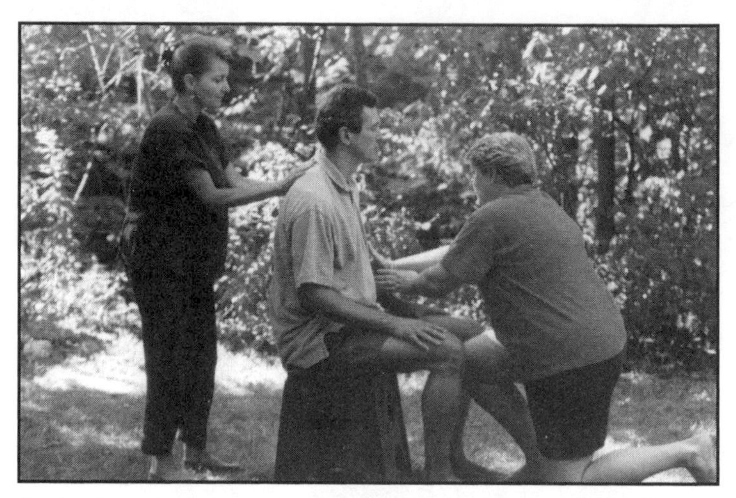

8. As Stacey continues her evaluation, Jan examines the healee's nuchal area for imbalances of flow or other cues. It is in this area, overlying the brachial plexus, that ancient texts state that the flows of the vital-energy field meet after replenishing the supply of prana at the various organs of the body.

9. The two TT therapists have switched places, each examining the other "side" of the healee's vital-energy field.

10. The two TT therapists keep rhythmic pace with one another in their assessment, moving down the body.

11. When Stacey arrives at the seated healee's hips, she maintains that position in support of her partner, while Jan continues her assessment down the healee's trunk and legs.

12. Often, questions and information pass silently between the two TT therapists during the session. This telepathic bond frequently remains strong between the two TT therapists for some time afterwards.

13. Stacey checks a hunch about the healee's lymphatic system. The healee is deeply into a relaxation response.

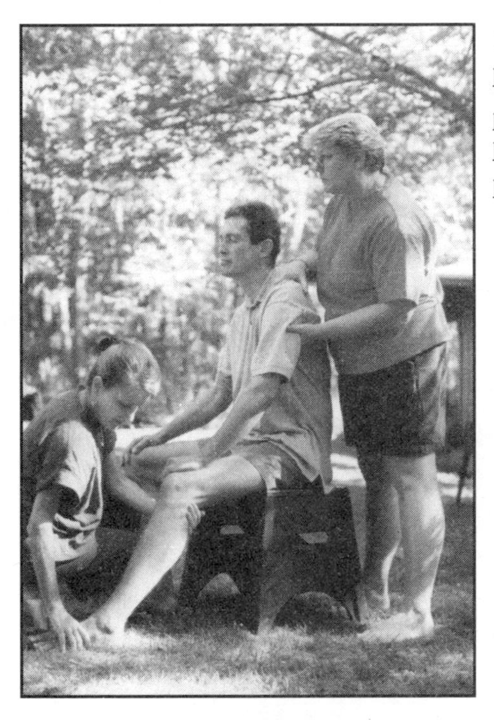

14. Stacey uses palpation to support her visualization of the healee's lymphatic glands. Jan is checking the healee's vital-energy flow down his leg and through the foot chakra.

15. Stacey is getting into position to send vital energy through the healee to Jan so that Jan can fully check the outflow of the energy through the healee's foot chakra. One can see from the position of Stacey's head that her attention is focused on sending energy through the healee, and Jan is in position to pick it up should the flow reach the foot chakra.

16. While Stacey gently projects vital energy from her hand chakras through the healee's solar plexus area (via the kidney area), Jan checks the healee's field in the area of the foot chakras to get a sense of the balance or imbalance of the vital-energy flows. The therapists are concluding the TT session.

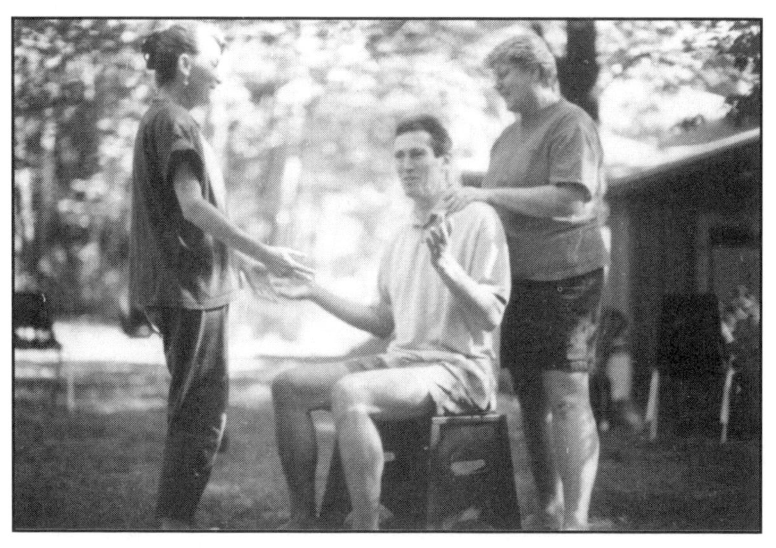

17. Friendly discussion at the end of the TT session.

18. The companionly repartee is friendly and fast, for the healers and healee know each other very well by now by virtue of the intimate nature of the Therapeutic Touch interactions.

19. Good relationships are apparent between healer and healee.

20–24. Examples of peer therapeutics. If a healee shows interest in Therapeutic Touch, the TT therapist will teach him the rudiments of this healing way. When the TT therapist is assured of his competence, she chooses a healee for the new TT therapist to work with. Ideally, the chosen healee will have the same illness or problems as the new TT therapist originally had, allowing the therapist to gain significant insight into his former illness while helping the healee. In several cases, such insight has led to a transformative and life-changing experience for the new TT therapist.

What Escapes the Camera

What did you see in the photos? Of course, an essential element is lost by viewing these immobile photos serially. One of the striking elements of the Therapeutic Touch enactment is the mindful, gentle, rhythmic movement that expresses the therapist's compassionate engagement in the healee's well-being. This movement, under control of the mind, reflects the analytical weighing of the many subtle factors that are revealing themselves to the therapist's awareness as she proceeds through the TT process. There is an elegance in the movements of these two therapists that is reminiscent of a stately dance, a pavane, where the measured movements of each reflect the other as the two quietly respond to subtle signals and move as one. However, in these photos the expression of their sensitivity is faster than the click of the camera.

The pattern of dynamic expression is one in which the TT therapists are expressing something greater than themselves, extending the boundary of self as they move into a larger reality. It is the translation of this happening that is so evasive of explicit description in common terms, and so one reverts to metaphor (see Table III).

Therapeutic Touch and the Pranic Flow

This discussion of Force B is an appropriate place to review the workings of prana and its flows because the five pranic subsystems that can be assimilated into the body at this time in our evolution play an important part in the healer-healee relationship during the TT process.

Prana is the major activator of the vital-energy field, the setting for Therapeutic Touch. Recall that prana is a universal principle, like gravity; fundamentally, it is the essence of all motion, force, and energy, and causes the vital movements by which we determine whether or not an object is alive. Paradoxically, prana is said to be in all forms of matter, but is not itself matter; it is breathed in with the air, but it is not oxygen; moreover, prana can penetrate where the air cannot reach.[4]

Table III. Metaphors as Translators, Bridges and Cues of the Inner World

balance	heat	pattern
blockage	high	pressure
centering	inner self	rebalance
cold	interiority	resonate
congestion	interpenetration	simultaneity
cues	listening	synchronicity
Deep Dee*	little electric shocks	Think blue!**
direct (vital-energy flow)	modulate (vital-energy flow)	tingle
field	multidimensionality	unruffling
harmony	order	visualization

*A term coined by a healee, Jean, to describe what it feels like when a TT therapist knowledgeably uses her chakras to deeply rebalance one's vital-energy field.

**A term that developed spontaneously among TT therapists and later came into popular use. Blue is a sedating, calming, or quieting color. In using it for an anxious healee, for instance, the TT therapist visualizes blue in a shade usually associated with the archetypal Mother of the World. The therapist tries to get a sense of what that shade of blue (perceived as light rather than pigment) "feels" like (in reference to its vibratory energy). She then projects both the visualization of the blue and the sense of its vibrations to the appropriate area in the healee's vital-energy field.

Prana is conceived of as entering the individual's body through the spleen chakra; however, there are several other avenues of entrance for the various subsystems of prana. A major percentage of prana reaches the body through the breath. Other subsystems of prana enter through the skin during Vitamin D absorption; through peristalsis and other rhythmic muscular movements that are largely initiated by the autonomic nervous system; through thought, as seen in mind-to-mind communication; and through food.

Prana makes itself evident through its ingenious organization of cyclic biological activity, which is foundational to the life process. Recall how vital energy streams through the core of the chakra and flows through a nonphysical stemlike shaft or stalk that enters the spinal cord.[5] Conveyed along the nadis, the vital energy flow travels throughout the body and revitalizes it, then returns to the chakras. It moves in spirals through the periphery of the chakras' spinning arms or petals, is refreshed, and then spirals out again in a constant rhythmic pulse of inflow and outflow. These energetic streams are then cycled in biorhythms, flow in tides, and are patterned by force fields.[6]

As this circulatory process of subtle energies proceeds, the spiral becomes increasingly wider and finally fades into the whole superphysical field of the person's chakra complex. The vital-energy flow appears to attenuate progressively and finally to vanish into the stuff of the universal fields, much as when one exhales, and the chemical elements in the breath diffuse into the atmosphere of earth. Upon inhalation, freshly charged molecules of oxygen and other gaseous elements that fuel the vitality of the life process are breathed into the body, and the vivacity of the person is strengthened.

Kunz writes of the importance of the characteristic rhythmicity of the chakras and their interrelation within the individual complex.[7] These relationships support a healthy state, whereas interferences or obstructions in the vital-energy flow tend to deform the normal, healthy energy patterns. These blocks can cause a loss of vitality and set the stage for illness.

Since the timing of the rhythmicity of each chakra is so exact, any interference with these individualized pulses can be visually discerned by people who are naturally proficient in nonphysical perception. It is in this way that psychosomatic illnesses can be recognized quickly by a few especially talented people; however, these rhythmic alterations can be perceived by many sensitive but non-clairvoyant TT therapists in various other ways. The therapist may sense a distinct change in the healee's vital-energy field through her hand chakras, or she may be aware of a felt shift in one of her own chakras in identification with the healee. She also may grasp the condition intuitively.

The TT Therapist as Test Object for the Universal Healing Field

The ability to remain unattached to the results of the TT session while retaining a deeply compassionate regard for the healee's well-being clearly marks the mature TT therapist as a useful test object for the universal healing field, whose chief characteristics are based in principles of affinity, alliance, and unconditional union. From my own experience, I believe that this pranic outworking of the universal healing field is reflected in the TT therapist's body language, gestures, and moods, which can be objectively observed by astute monitors. There are several indicators for this state of consciousness, most apparently:

- Mindful recognition of one's relationship with the universal healing field. This is most often seen in the TT therapist's growing awareness of the principles of order that underlie the healing process.
- The expression of compassion as a natural concomitant of one's worldview, and the willingness to be vulnerable to the effects of compassion on one's lifestyle
- The ability to focus with confidence built on knowledgeable intentionality in the direction and modulation of the flows of the several subsystems of prana
- An objective perspective of one's own ego strength and recognition of personal motivation

- Ability to focus on the healee's potentialities in reference to his wholeness or potential ability to act in an integrated fashion
- Willingness to act as a model or mentor as the healee learns life-affirmative styles and puts them to therapeutic use
- Receptivity, impressionability, and responsiveness to others' vital-energy fields, particularly those in need. It is this flexibility and ease of mobility in the vital-energy field that allows the TT therapist to resonate so reliably to the healee's need, and fosters the ability to be "present" to deeper states of consciousness. Because these abilities are integral to her function as healer, it is important for her to "exercise" her own vital-energy field and keep it pliable, supple and accessible. Several exercises suggest themselves:
    - Spontaneous acts of compassion, love, empathy, kindness
    - Considered (mindful) projections to others of thoughts of peace, calm, gentleness
    - Meditation, particularly in reference to visualizations that increase sensitivity

Please see Table IV (next page) for indicators of universal healing field activities, and Chapter XIV for additional exercises.

The Healee's Energy Fields

Vital energies, as we have said, are in ceaseless motion. They are continually rearranging patterns, modifying rhythms, and transposing biochemical and psychodynamic networks to refashion templates of form, function, and behavior that nevertheless remain unique to the individual. There are several physiological signs in the healee that are

Table IV. Major Indicators of a Responsive Vital-Energy Field

accessibility	graciousness	relaxation
adaptability	imperturbability	receptivity
affection	impressionableness	self-sacrifice
appreciation	insightfulness	sensitivity
aptness	kindness	suggestibility
calmness	limberness	suppleness
charitableness	litheness	sympathy
concern	lovingness	tenderness
compassion	mercifulness	thoughtfulness
empathy	moldability	tolerance
flexibility	non-rigidity	understanding
forgiveness	perceptiveness	unselfishness
friendliness	pliancy	

indicative of significant change taking place in his vital-energy field. One of the early signs of stress, a most influential precursor of vital-energy imbalance, is muscle tightness behind the neck. Here, the various subsystems of prana meet in the field of the brachial plexus (situated between the shoulder blades directly under the nuchal muscles) after they have circuited throughout the body. The TT therapist pays particular attention to this sign by working over that site at the base of the neck (as well as over the anterior neck area to get at the throat [vishuddha] chakra). Another sign at the throat itself is a sense of fullness that may progress to feelings of outright pain. This most often occurs in persons in grief, bereavement, depression, or other deeply emotional, unstable states.

The site of the solar plexus chakra is also very sensitive to change in the psychodynamic field—more so, in fact, than in the vital-energy field—though of course the fields interpenetrate and are highly responsive to one another. The major physiological sign of disturbance in the psychodynamic field is a sense of queasiness in the physiological solar plexus area, or a feeling that the solar plexus has "flipped" or turned over, which may be quickly translated to feelings of nausea or dizziness. Another sign, particularly in cases of fear, agitation, despair, phobias, or other signs of emotional disquietude, is change in breathing rhythms.

Breath is one of the most sensitive indicators of change in these fields. Since breathing is a primary source of prana for the body, changes in breathing patterns sensitively register biochemically and physiologically the state of our vital fields. Other signs that the TT therapist pays close attention to are flushing, pallor, and the healee's inability to focus attention. Most of these signs are largely of autonomic nervous system origin, and so it would benefit the TT therapist to become knowledgeable of this system, with particular attention to recent psychopharmacology studies.

Attitudes and Behaviors That Affect the Vital-Energy and Psychodynamic Fields

There are several attitudes and behaviors that significantly affect both the vital-energy field and the psychodynamic field; the physical structure of the body then mirrors these influences both biochemically and neurophysiologically. If the attitudinal changes persist, they tend to form deeply grooved field patterns and vortices that display themselves as ingrained characteristics of the personality: attitudes of acceptance or of anxiety; acts of centeredness, compassion, confidence, intentionality, and love; or feelings of restlessness, fear, resentment, or depression, etc. All merge out of strong field patterns that influence the sculpting of the personality throughout life. Since these attitudes and behaviors have significant relationships to the solar plexus chakra, they act by regulating the flow of vital energies to the physiological analogue of that chakra, the adrenal glands, and depress or stimulate their functions as well as that of the vagus nerve, which richly feeds into the area, e.g., the heart, lungs, and viscera.

Negative attitudes and behaviors (especially resentment, which seems to be a particularly virulent emotional contagion of our time) exact severe energy drains on the individual. If left unchecked, such counteractive emotions can cause physiological damage and in some cases (e.g., peptic ulcer) anatomical injury, mutilation, or loss. The best way to treat healees with such personality traits is to anticipate the probable sequelae of their actions and to help them learn new, preventive, life-affirmative lifestyles.

With the help of Dora Kunz's remarkable insight, over the years several of the subtle effects of negativism during TT sessions have become apparent to us. Of the nugatory emotions, our experience has indicated that depression is the most dangerous; it not only imprisons the depressed person in the rigidity of his own vital-energy field, but also affects other persons in the vicinity who are vulnerable to negative, pessimistic, or melancholic thoughts. The most pernicious of the negative forces, depression literally imprisons or congeals the attention of the depressed person to a confined area of his psychodynamic field, which then acts as a formidable barrier to interaction with outside forces. It is very difficult for the TT therapist to break through this barrier for therapeutic purposes.

The most effective force that I have found is the strong, focused projection of love to the healee;[8] however, to conserve the healing effect so that it will continue to work in the healee's life, the healee himself must deeply desire to change to a more life-affirmative style of living. (See the exercises on pp. 147–148).

Several other effects we have noticed are that high anxiety inhibits a person's ability to "reach out," literally to stretch his vital-energy field; in fact, it appears to the TT therapist that the anxious healee is thrown back onto his own subtle energy resources, the vital-energy flow forming a closed loop upon itself. This self-induced backlash of anxiety and fear can sharply escalate the healee's emotional state to the edge of panic. In the case of acceptance of one's illness, however, there is an unmistakable sense of disengagement by the healee, of "letting go" of the results of the interaction—that is, of being attached or locked into a specific outcome—and the energy flows return to a more normal pattern of vital streaming.

The most important factor disclosed during the healing interactions was the willingness of the healees to accept change in their lives. Once this acceptance had been declared (verbally or in mindset), we saw a commensurate shift in the healee's mood and a decided relaxation response. It is the latter that allows therapeutic access to the healee's vital-energy field.

To Change the Quality of the Vital-Energy Field

TT therapists use various techniques to help balance the subtle energy fields, including visualization, meditation, intentionality, inner quietude, prayer, confidence, and love. The following suggestions can also change the quality of the vital-energy field, and are useful not only to the TT therapist, but to the healee seeking lifestyle change. For instance, we have learned from Dora Kunz's detailed reports on subtle energy dynamics that it is particularly crucial for depressed persons to be physically active (her favorite prescription for them was to take up rollerblading).

- During daylight hours, work or play outdoors, preferably in a non-polluting area, for one to two hours, three to four days a week.
- Spend one to two hours, two to three days a week, in aerobic activity at a health club or gym, or at home.
- Make it a habit to do unobtrusive abdominal breathing exercises by pursing the lips on exhalation, in spare moments and

especially in times of stress. Within a short time it will invoke a natural relaxation response.

- Swim, shower vigorously, use a hot tub or a jacuzzi as often per week as your schedule permits. Rub your skin with a pumice rock, Japanese bathing brush, oatmeal, or other skin scrub.
- Visualize a place that you enjoy going to, e.g., a favorite or admirable tree. When you are "there," be present to the tree: try to "feel" the texture of the bark of the tree, "hear" the sound of its leaves rustling in the wind, "watch" its branches swaying, "talk" to it through mind-to-mind communication techniques. If it answers, consider yourself lucky!
- Eat nutritive meals when you are hungry. If you are not hungry, drink pure water plentifully and often.
- Intend to sleep before midnight.
- When outdoors, take off your shoes and walk barefoot, particularly if you are walking in clean grass, on a forest floor, or along an ocean shore.
- When you are tired of walking, dance!

Fine-Tuning the TT Assessment

During the TT assessment, the therapist's consciousness is concentrated on indications of order, pattern, flow, rhythmicity, symmetry and balance, or their adverse indications: random disruptions of functions, irregularity and hyper- or hypoactivity of the vital-energy flows, congestion or pressure, dysrhythmias and arrhythmias, imbalance, and asymmetry.

As the therapist goes on to rebalance the healee's fields, she notices that the functions of her two hands are not the same. The dominant hand directs and modulates the healing energies being transferred to the healee, while the other hand serves to "anchor" the hand chakra in the vital-energy field of the healee and to "listen" for indications of imbalance. However, these two avenues of information—the left and right hand chakras—work closely together to bring to the therapist's mind a clear picture of what is happening. For instance, a sense of "tingling" in the TT therapist's hand signals to her that there is a lack of energy in the healee's fields. An irregular pulsation of the vital-energy field might indicate a hormonal imbalance. It is vital that the therapist be able, while deeply centering her consciousness, to attend to the appropriate techniques, the healee's frame of mind, etc.—to be aware of the healee as a total being.

Dora Kunz has reported a stretching of the therapist's vital-energy field during the TT session. The healer experiences a breakdown of restrictive patterns within the field and a sharpened feeling of unity, because the energies of the vital-energy field and the psychodynamic field are loosened, permitting flow and repatterning of these energies into associated linkages or networks. These patterns in the individual's fields are what we call attributes, habit patterns, emotions, etc. This phenomenon, Kunz explains, lends another dimension to how our relationships with others are perceived. For the persons who are trying to realign their lives in order to better help or heal those in need, she emphasizes that the effort to break one's own emotional patterns demands persistence and consistency of effort, that it cannot be half-hearted or sporadic. Rather, the decision for self-change must be one's primary concern, overriding other interests and desires.[9]

Influences to Be Taken into Account

Therapeutic Touch is by no means an easy path. Dealing as it does with so much that is rooted in the psychic world, it is difficult to maintain the needed sensitivity to people's emotional and physical needs while keeping one's own ego stable and one's decisions and recommendations practical. Studies in the various fields of healing have sounded warnings we would all do well to heed. For example, it has been indicated that during the assessment phase of the healing session, which is the prime basis for the selection of the strategies that will be used, the healers can easily be influenced by:

- Suggestion and persuasion
- The order in which cues are picked up
- The way in which the information the healer receives is framed
- Recent experiences of the healer
- Distractions

Data have shown that as a group, healers are not good at considering multiple factors, for we may give some factors too much weight, ignoring others out of unrecognized bias. In developing Therapeutic Touch we have tried to keep such precautions in the forefront of our attention.

- Staying in a centered state of consciousness throughout the TT process is critical to its success as a transpersonal means of healing. If you go off center while engaged in TT, make sure the healee is comfortable and in a safe position, and stop. Try to recenter. If this is not possible, refer the healee to someone else who can help him.
- Do not try to rationalize your findings during the assessment phase. Accept the conclusions of that state of consciousness, act on those decisions, and then test their verity by the results of the TT session on the healee's condition.
- Therapeutic Touch should not cause you to sweat. Energy flow is natural and nonphysical. Its direction or modulation originates in the mind and cannnot be "squeezed" out. Instead, it is empowered by the TT therapist's intentionality, coupled with an effortless (e.g., unstressful) employment of the therapeutic use of self.

- Use Therapeutic Touch particularly gently and sensitively with:
  - babies
  - people with head injuries
  - pregnant women
  - highly traumatic cases

(A useful rule of thumb, I find, is that the more injured, sickened, shocked, confused, or frightened the healee, the more sensitive, gentle, and aware I try to be. Invariably, you can return, observe the effects of the TT treatment on the healee, and then decide whether further intervention is warranted.)

- Rhythmicity is of prime importance in the TT rebalancing phase, particularly between two TT therapists working on the same healee. If each works at her own pace without regard to what her partner is doing, it can be very traumatic to the healee's vital-energy field and may cause dire consequences.
- Do not become attached to the results of your efforts. The healing process is very complex and, as we are seeing in this section of the book, its full unfolding depends on factors we may not understand.

The Healing Session

As the therapist is the test object for the universal healing field, so the healee is the prime test object for the effectiveness of the healing session. With this in mind, we must also recognize our ignorance of the effect of the healing milieu on a particular healee.

The ideal progression of the healee begins with the relaxation response, which will occur within two to four minutes. The healee feels this as a noticeable reduction in tension. The vital-energy flows resume normal movement, and with the sensed return toward normalcy, the healee feels a sense of exhilaration that can develop into a continued state of well-being. There is a concomitant decrease of fear and apprehension and a lessening of anxiety, which affects the immune system favorably.

Dora Kunz has described this occurrence thus: the continued stream of vital energies directed to the healee by the TT therapist causes the healee's vital-energy field to expand. This action neutralizes or removes some of the disturbances or blockages to the periphery of the healee's vital-energy field, resulting in a reduction of anxiety and increased stability of his vital-energy field. The immunological system is strengthened by the continued flow of vital-energy, and the underlying rhythmicity of the pranic flow becomes more coordinated, producing a general relaxation response. This response facilitates the ensuing healing process to the extent that the healee accepts this progressive change toward wellness. In some cases, often because of secondary gains, people return to the "habit" of illness. However, under the best of conditions the healee's inner self takes over and the healing progresses rapidly; in a few cases this can occur spontaneously.

We can begin to realize the complexity of the situation if we examine a healee with a common ailment, such as pain. When a person is in pain, he naturally turns inward. In doing so, he very frequently will unconsciously create a nonphysical barrier between himself and his environment. This cocooning not only isolates the healee but acts to reduce his vitality. This state of low energy increases the healee's vulnerability to several uncomfortable symptoms, such as:

abdominal tightness
depressed breathing rhythm
depression
immobilization
indecision
irregular heartbeat
irritability
lethargy
moodiness
nausea
oversensitivity
restlessness
sense of exhaustion
tension
withdrawal

Any group of symptoms in this amalgam could be indicative of an illness or an iatrogenic effect (a side effect of a prescribed medication). The TT therapist should be aware of both possibilities, for the healee himself is often most bewildered by multiple and related sources of pain.[10] This confusion can serve to magnify the problem, although pain itself responds to Therapeutic Touch very well. Several of these symptoms taken together might indicate unexpected or serious complexities nested in the healee's problem that may be beyond the healer's expertise; in such a situation, it is wise to refer the healee for further appropriate professional advice and treatment.

Force C: Forms and Intelligences of Nature

Sensitivity to Nature and Place

It is striking that ethnic peoples everywhere in the world "participate in the landscape"; that is, they sanctify the local features of the land.[11] Interestingly, when we do Therapeutic Touch outdoors we, too, become very sensitive to the living qualities of the environment, whether they be flora, fauna, or geological features—even climate and the celestial bodies. These are part of the exquisite cosmic order of which the TT therapist becomes aware through the act of healing.

Among native peoples, sites that have a particularly strong energetic atmosphere are considered sacred places. As the idea of sacred places has become better understood in recent years, the concept has been integrated with that of deep ecology.[12] Joseph Campbell has said that sacred places are "where you can simply experience and bring forth what you are and what you might be." He goes on to encourage people to seek out sacred places because "people claim the land by creating sacred sites . . . they invest the land with spiritual powers."[13] He relates his own youthful experiences in a forest, observing great owls and other wild animals, and describes their effect on him thus: "All these things are around you as presences, representing forces and powers and magical possibilities of life that are not yours and yet are all part of life, and that opens it out to you. Then you find it echoing in yourself, because you are nature."

In China, geomancy has developed as an art that reveals strong life forces (ch'i). These are associated with specific sacred places in mountains and valleys, caves and caverns, the vicinity of certain stones, or near springs, waterfalls, streams, or lakes. Swan talks about the Salish of the Northwest and their traditional places of power or presence. A sacred state of mind can develop in such atmosphere "where magic and beauty are everywhere." He calls this Salish state of mind "skalalitude."[14]

The concept of sacred place in nature is so deeply embedded in Native American traditions that it is listed as an official category in the American Indian Religious Freedom Act of 1979. However, this sensitivity goes beyond the confines of North America. In France, almost 150 years ago, a young girl had visions of the Virgin Mary in a sacred place, a grotto in Lourdes. Her experience has given hope to hundreds of thousands of sick and disabled people who have flocked to petition Our Lady of Lourdes for help. By official count, only one percent of them have been "cured," but the experience has changed the quality of life for many, many more who were healed, often as a result of transpersonal experience.

About twenty years ago, I found myself in a sacred place, a cave at Uluru, almost in the center of Australia, with a mixed group of park rangers (some of whom were Aboriginal people), sharing the creation stories of my adopted Apache family—which were incredibly similar to their own stories. We also discussed similarities and differences in the two cultures' ideas about sacred places for several hours, long enough for the molten sun's heat to cool and a brisk, freshening breeze to rise. I remember leaving that dreaming haven of the Rainbow Snake, Wanambi, and realizing that the shift in shadows on Uluru as the sun was going down brought out a very different aspect of that sacred place from the features I had seen when I went into the cave; the site had an unusual magnetic quality to it that has stayed with me through the years. Physicists say that the quality derives from a telluric energy whose current runs through the rocks—a similar energy to that used to set the alignments for the Mayan and the Egyptian pyramids, Chartres cathedral in France, the Taj Mahal in India, and many other notable structures throughout the world.

An Experience in a Stone Circle

An assumption of Therapeutic Touch is that to understand its process one must be willing to use other senses than the half-dozen our society most easily recognizes. For instance, the TT therapist's centering induces a state of consciousness that enhances interior experiences, including the realization of personal knowledge. Under the best of conditions, this practice serves as a check on the reliability of one's intuition, as well as on the accuracy of the TT assessment. When one attempts to recreate in the mind the actual experience of the assessment, it is possible to recall the state of consciousness prior to the intuitive flash, or to remember a feeling tone that served as a cue to that state of mind. Thus, with continued and conscious practice, the TT therapist begins to gain an understanding of the preparatory conditions that precede and facilitate the intuitive state. Tentatively at first, and then more frequently as repeated trials test out the reliability of such "knowledge . . . without preparation,"[15] the therapist accepts this higher-order structure and puts it to the service of healing.

Consistent encouragement and continuous testing of the use of intuition in the process of healing may stimulate the unfolding of confidence in this higher-order resource of the self. Over time, and after numerous subjective testings of one's intuitive grasp, a sense of certitude may intensify, so that the therapist becomes willing to trust life decisions to its insistent urgings. Even though the promptings of intuition may seem unreasonable, irrational, or even contrary to the accepted reality of the day, they demand recognition because they have proved their validity. This inner "listening" becomes in time an innate expression of personal lifestyle. This can be of critical importance in times of crisis, when one "listens" very carefully, as if to a deeply respected friend.

These higher-order resources provoke experiences in interiority that penetrate into the matrices of one's everyday lifeway as experience in the TT process is deepened. For this reason a personal experience may serve to illustrate the degree to which one can develop seemingly unreasonable trust in strong intuitive perceptions that, since they speak to sacred places and intelligences of nature, will prove on target.

A few years ago, I was invited to London by a private foundation to present findings on research I had done on the TT process, and to teach some of its techniques. The schedule permitted me two days to overcome my jet lag, and I was asked how I would like to spend the time. One of my choices was to visit some ancient stone circles other than Stonehenge, which I had heard was quite crowded with tourists. Happily, a friend of the Foundation owned a piece of property in Oxfordshire that contained another three-thousand-year-old stone circle, called Rollright. The Foundation's Director of Research and I were granted permission to visit Rollright, and we planned to be on site in time for sunrise. Very shortly after our return to London, I wrote the following account of our visit, which has been minimally edited or changed in any way:

Ruth picked me up a little after 4 a.m. and without traffic we drove up-country quickly to the private land that is the site for the Rollright stone circle near Oxford . . . Ruth was a very thoughtful companion and gave me both the space and the time to do my own thing until finally the raw, wet cold driven by a penetrating wind forced us to leave and seek out a homey and delightfully cozy, warm inn for breakfast. While at Rollright, I felt that I had a quite meaningful experience with the stones and so, with the intention of checking with knowledgeable friends when I get back to the United States, I decided to record this very subjective impression while it is still fresh in my mind.

Since the ground was quite muddy, I sat on a low stone directly opposite two comparatively tall stones, between which I could see the sun beginning to rise. The site was a good choice, as the time of year was close to the vernal equinox and all of nature seemed vibrant and alive. The stones themselves are massive, darkly colored and irregularly shaped, looking like large gnomes. They are not straight-sheared as are those at Stonehenge, nor are they of similar material. These stones have many holes in them, seem of volcanic origin, and appear to be deeply rooted in and at one with Earth itself. Wondering how to get in touch with such an ancient structure, I decided to center myself and then allow impressions to arise to consciousness while I meditated.

The impressions I had were not at all what I think I expected. At one point in my meditation, I became aware of an energetic interchange among the stones, actually between the stones, the earth itself, and the constellations of stars or other celestial bodies then in a particular geometric (astronomical?) relationship to one another.

The energy flow was visualized in two distinct styles: one flow concerned itself with the stones themselves and seemed to flow around the circle and through the stones in a continuing cycling of energy. The second flow was quite different. For one thing, its flow was linear rather than circular, so that it formed a grid or latticework as it flowed across the circle floor in east-west, north-south directions. These linear flows formed geometric patterns of triangles, squares, or rectangles, all of the geometric patterns interconnected and of one piece. The energy flow seemed the color of white gold, and the inside spaces of the triangles and rectangles were of a color I'd describe as black, although that is not precisely what I have in mind.

As I "watched," the impression I got was that the patterned energy flow (I'm not quite sure about the circular flow) was a spin-off or a product of a dynamic interaction between the stones, planetary energy coming from Earth itself, and certain constellations then in the sky. The stones, I sensed, acted as transducers for the energy streaming down from these constellations, which was then used in some way by Earth. My impression was that elementals (?)/gnomes (?)/ the stone-life (?) were involved in this process, and the whole transaction was an important part of the life (i.e., 'behavior') of these stones. As (when?) the laying down of this carpeting of geometric energetic patterning completed itself, there seemed to be a slight tremor that occurred throughout the structure (which, strangely, was more rectangular than square, so that the energetic structure went beyond the actual stone circle itself), and then the whole energetic structure quietly raised itself as one vibrating piece from the ground, a white-gold shimmering latticework, and effortlessly floated directly upward from the Earth's surface into the sky. At some point it rapidly disappeared, but I didn't see it do so. By this time my eyes shot open in startled surprise, but of course I didn't see anything then except the as-usual English countryside blowing in the sharp wind, noting that the sun now was beginning to go behind some scud clouds.

Many associations immediately flooded my mind, but the predominant one concerned a comparison between the geometric patterning I had perceived and descriptions in various religious works of the tesselated floor of subtle energy evoked by various church ceremonies. On the heels of that association came a very clear thought that human ceremonials (with people) would enhance the effect I had visualized, if they were done in conjunction or in cooperation with this natural enactment. In the meantime (or, perhaps, in the times when no people took part) these natural forces would diligently continue their never-ending ritual.

As we left the area, this sense of continuance was reinforced by my last glimpse of the stones as the car sped away: the stones, still looking like giant, hulking gnomes, seemed to be deeply engrossed in the production of another latticework of subtle energies. One had a sense of complete concentration on the task at hand, which would go on whether or not we or other people were there. I noticed that although I was relaxed in the car, in addition to a good sense of well-being, I felt highly alert and very aware. I realized that I felt a sense of accomplishment, a sense that the impressions I had experienced in some way fulfilled the reason for my coming to England in the first place, even though I had not realized that I had had any 'reasons' to begin with!

The Dragon Project Report

Having committed these impressions to a black-on-white statement that I could confirm or reject upon my return to the United States, I put the matter aside and went to a small dinner party. In the rush of subsequent events, I forgot about the incident until two days later, as I was greeting guests in a reception line at the Royal Society of Great Britain in London, having just delivered my paper.

Toward the end of the reception, a young lady approached me, and said, "Knowing that I was coming to hear your paper and that you had visited Rollright the other day, Don asked me to tell you that he'd be especially interested to know your impressions."

Surprised by the directness of her statement, and wondering who Don might be, I said, somewhat equivocally, "It was only an impression." Nevertheless, I went on to tell her the story of my visit.

When I finished, she said, "Oh, I see that you know of our Dragon Project."

"Dragon Project? What do you mean by Dragon Project?" I asked, not at all sure whether to take her remark seriously.

"Don has been studying the Rollright stone circle for some time, and the study has been named the Dragon Project." She continued, "I will see that you get a copy of his latest report."

We parted after that, and a day or so later I received the promised report. By that time, I was preparing to return to the United

States, other priorities vied for my attention, and I put the report in my briefcase.

It was about midway on the return flight that I opened my briefcase and inadvertently came upon the Dragon Project report. Curious, I took it out and flipped through the pages. As I scanned the report, sentence after sentence seemed to leap out at me. I felt the hairs on my head rise, and my scalp actually prickled in sympathetic reaction to my surprise. Several statements on the findings of this high-tech empirical study seemed to corroborate my subjective intuitive impressions. Now fully intrigued, I returned to the beginning of the Project report, carefully read it in its entirety, and mused on its relationship to my own experience.

The Dragon Project involved a study of "the nature of stone and the possible routes of energy storage and conversion." "Don" was Dr. G. V. Robins, at that time a Fellow at the University of London and chief coordinator for physical monitoring at the Rollright stone circle site. The basic assumption underlying the study was that at the molecular level the structure of stone is an incomplete silicate lattice that contains a shifting population of electrons, which are trapped in imperfections in the lattice or in the microcrystalline cavities. These electrons are in continual flux as they escape from the lattice traps and migrate through the stone, the report says. Due to this emigration, there is a small but measurable drifting potential that, they think, probably drifts to earth in a standing stone. Because of this, "the role of stone in any manifestation of Earth energy might be in producing electrical phenomena."

Such potential current conveyors would only form a minimal fraction of background "noise," so an effort was made to enhance their absorption of suitable electromagnetic energy (microwave radiation is such a suitable energy). To rule out distortion in the silicate lattice from trace metal impurities and magnetic field variation coming from local geographical sources, they decided to monitor the transduction by the stone lattice of incident microwaves to ultrasonic microwaves. They monitored the ultrasonic effects at the Rollright stone circle, known as The King's Men, as well as the nearby lone "King Stone" and the adjacent collapsed dolmen known as the "Whispering Knights." The monitoring also included a wide radius of the surrounding countryside. The following findings seemed to dovetail with my own impressions:

1. The levels of ultrasonic intensity followed a distinct bimodal distribution, with the greatest intensity occurring around the time of the equinoxes and the least intensity occurring around the period of the solstices. At the vernal equinox, it was found that the stone circle activity significantly increased, whereas the activity at the King diminished. (The activity of the Whispering Knights closely followed that of the King Stone on all occasions.)
2. There was a significant differential of activity noted inside the circle, between the stones in the circle, and on the stones themselves.
3. The complex waveforms of the periodic pulsations of the ultrasonics indicated that the site activity "continued for the same length of time at all parts and finished abruptly on all occasions."
4. There was "a clear and precise cut-off point, usually a few feet outside the circle."
5. On one occasion, the monitor recorded that as the generalized site activity lessened, the pulsing around the ring of stones nevertheless continued "in the narrow corridor of the perimeter" of the stone circle, somewhat similar "to a cyclotron effect."

The results of the ultrasonic monitoring suggested to the investigators that the stone circle could be considered "a three-dimensional dielectric antenna whose orientation allows maximum energy transduction at the time of the equinoxes." The investigators also put forward two possible uses that those ancient builders, people of the Bronze Age, might have had for the site: the healing effect that is associated with weak electrical fields, and the increased grain germination rate that occurs within certain ultrasonic frequencies.

The possibility that the Rollright stone circle might be a healing site drew my attention because a similar function was once attributed to Stonehenge. In a twelfth-century account by Geoffrey of Monmouth, Merlin says of Stonehenge, ". . . in these stones is a mystery and a healing virtue against many ailments."

As I came to the end of the report, I was amused that my own experience at Rollright so clearly supported The Dragon Project investigators' concluding remarks. After discussing various possibilities for future research, they write: "Of one thing we can be sure, that whatever [further information] we generate, we will continue to be surprised by it." And I found myself responding: Verily, verily . . .

Force D: Natural Powers of Therapeutic Intervention

Impressions of Angelic Presence

As it turned out, only Padre and I believed in angels. I was then a student nurse, and Padre was a staid Episcopalian minister who had taken on our class and formed it into a fairly presentable choral group. We were gathered around the piano and had been talking about the hallowed atmosphere in many hospitals during Christmas and Easter. As later became my custom, I had volunteered to work at the hospital during Christmas and Easter that year, and that decision had provoked the discussion. Both Padre and I were deluged with questions about our assumptions about angels: How did we know "they" were angels? What did they look like? Did we see the angels inside our mind, or were they standing nearby? If they spoke, what did they say? and many more detailed questions. Some of the questions even sounded angry. Padre, in his quiet, easy manner, thought their response was funny, made light of the questioners' comportment, and assigned them various texts in the Bible.

Padre never told us whether or not he saw "them." I, of course, cannot see them, and could only report on the feelings that arose within me on the occasions when I thought angels—or other beings, I wasn't choosy—were present. However, those impressions—occurring most often when I am out in natural surroundings, but also frequently when I have been working as a nurse in hospitals—have remained constant over the years, and I can easily recall their characteristic details.

The experience of angelic presence is always uplifting and expansive. I have a sense of a conscious presence, but I can tell that the consciousness and sensibility are somehow different from ours. I feel a strong, single-minded, focused force that is protecting, loving, inspiring, and serene. I don't have a sense of a form, but of penetrating, lustrous eyes shining with compassion, and a far-flung radiance that has a fiery, opalescent quality to it. Usually this being is centered on some individual or group of individuals who are in sorrow or in pain. Very often, the person who is the focal point of this display is dying or very sick. I am not always aware of the results of these interventions, but invariably the people involved seemed to be comforted. Frequently, there is an attitudinal change about some critical issue that becomes apparent in their subsequent behavior.

In reference to Therapeutic Touch, I cannot say I have seen many miracle cures, but there is a wide range of TT therapists in the United States and abroad whose healees have experienced unexpected outcomes. My colleague, Dora Kunz, with whom I founded and developed Therapeutic Touch, was the fifth generation in her family to be born with the ability to see subtle energies. Through disciplined effort, she honed this ability to a fine edge. She wrote an outstanding book on her experiences with fairies, which included some material on angelic presences as well.[16] Shortly after the publication of that book, I asked her a question about angels. Her reply was not what I expected. Looking into the distance, she said, "Unfortunately, people will more readily believe in fairies than in angels." Perhaps that is where the question of whether angels exist must lie at this moment. For myself, I will reaffirm that I do think they will respond to sincere petition, and I do think they are very good at what they do, whatever that may be.

My Friends, the Trees

However, I *can* talk about trees, and the consciousness that enfolds them, from personal experience. I have always found comfort, communion, and a sense of protection in trees no matter where I've been in the world. Trees were my first friends, and we have a long history together, some of which relates to Therapeutic Touch. One such story, presented below, illustrates the increased sensitivity of the TT therapist over time; another may provide useful information about using chakras to communicate during Therapeutic Touch sessions.

Talking to trees (and having them answer!) may seem quite farfetched; however, there are many utterly pragmatic situations in which a person can prove to herself, if not to others, that there is validity in intentional exchanges between oneself and nonhuman consciousness. Communicating with pets, of course, comes readily to mind—the unique intelligence of horses, for instances, prints indelibly on the mind of those gifted with opportunities for such communication. For the healer, the ability to communicate with nonhuman beings is not an end in itself, but arises as a natural concomitant of increased sensitivity to all that is living.

How the Tree Found the Dog

A personal incident that is still very vivid in my own mind may serve to illustrate the heightened sensitivity of the experienced TT therapist. A friend had flown with a young red setter dog directly from Paris, France to Pumpkin Hollow in New England, where Dora and I had done the original research and development of Therapeutic Touch. During the night, the dog was let out and did not return, and so in the morning the fifty or so people at the Hollow formed search parties and drove off in their cars to find her. I did not own a car at that time, and so I set off with my own dog to search the meadows and backwoods.

It was a very hot summer day, and after an hour or so of search I sought out the cool shade of a large and very old maple tree. I sat with my back against its trunk and idly watched my dog sniffing out the territory. I thought to try to get in touch with the maple tree, and so I centered myself in what I thought was the appropriate chakra to communicate with this tree. I visualized the lost dog as clearly as I could and asked the tree if he knew where I could find her. I was aware of a response within myself almost immediately: in lucid terms, he told me to go toward a well-known lake and to ask the people as I went. Somehow the message instilled a sense of urgency in me. I quickly whistled up my dog and took off in the direction of the lake without questioning the instructions.

I cut across the meadow with a sense of purpose until I got to the lake road. I then stopped at every farmhouse on the way to find out whether anyone had seen the dog, since this was my interpretation of the instruction from the tree. I finally arrived at the lake without having turned up a clue to the whereabouts of the dog.

By now it was late in the afternoon, and as I looked at my tired pooch I thought we had been following a will o' the wisp that had originated as a figment of my own imagination. There was a free-standing telephone booth at the lake shore, and I phoned the Hollow and asked a friend to drive over to the lake and pick us up. It was then that the miracle happened. As I looked up from the phone, I saw a middle-aged couple, evidently tourists, walking along the lake's shoreline. Remembering the tree's advice to ask people as I walked to the lake, I went forward (with some ambivalence, I admit) to meet them and ask about the lost dog. Incredulous as it may seem, the wife turned to her husband and said, "Dear, wasn't that a red setter that was with the gateman when we drove back to the hotel last night?" The husband agreed, and we all walked over the nearby isthmus of land to a hotel situated on a bit of isle in the lake.

The husband sought out the hotel manager, who brought the gateman to us. Yes, he said, the dog we described had come to the gatehouse around 11 p.m., looking as if she had been running. He had given her some cool milk, but regretted that hotel policy would not allow him to keep the dog. When around midnight one of the chambermaids was driven back to the hotel by her brother-in-law, the brother-in-law noticed the dog and offered to give her a good home. The gateman gave the dog to the brother-in-law, who drove off with her.

That was all the gateman could tell us, but he did seek out the chambermaid at our request. She confirmed the story and told us that the dog had been taken to a very large town about fifty miles from where we then were. She gave me her brother-in-law's address and phone number and—thanks to quite explicit advice from an old maple tree—dog and mistress were soon reunited.

The Meaning Comes through the Experience

One can look upon this incident from other points of view than the one I suggest—the explicit advice of an old maple tree. However, one can indeed learn from these experiences. Joseph Needleman says, ". . . the real human learning involves an exchange of energies and an interaction of forces identical in nature to the movement of forces by which the universe itself comes into being." One accepts its validity, he says, because the experiences "resonate within the interstices" of the grid of events of one's life.[17]

In healing, much of this meaning comes through the process itself—the experiential, subjective search for personal values and beliefs. The interplay of inner events comes to conscious awareness through the media of metaphor and intuition, subliminal cues and paraconscious cognition, subtle physiological changes and deep gut feelings. There are unfettered primal energies that may thrust themselves to the forefront of the too-frequently censored and circumscribed focus of our immediate attention. They force the recognition that we are, indeed, of a similar nature as the universe.

The Intentional Use of Chakras

The second experience I mentioned above concerns the use of chakras during Therapeutic Touch. For a sensitive healer, these subtle energy foci can become conscious functional centers of awareness, sometimes at very profound levels.

My own use of chakras in healing is varied; however, when I want to assess the level of a healee's illness or trauma, I gently attune my awareness to my own chakras in an ascending fashion, so that I go into successively deeper states of consciousness. Simultaneously, I try to sense whether I can maintain communication with the healee on these levels; that is, I try to determine whether an active "line" of communication exists between us at what I believe is a particular chakra level. At some point a limit is reached, which I recognize as either my own circumscribed abilities or the healee's limitations. I then try to maintain this level of conscious communication between our chakras during the healing act. This is essentially what has come to be called "the Deep Dee," a term coined by one of the healees, a former journalist, with whom I have worked.

This intentional use of the chakras for therapeutic purposes is an enriching experience that enhances one's knowledge of self and others. It also may open to the TT therapist non-ordinary levels of communication; therefore, it becomes the responsibility of the individual to exercise intelligent discretion.

The Mindful Use of Chakras as a Lifestyle

In my own experience I have noticed a concomitant quickening of understanding of their own chakras in students as they begin to make Therapeutic Touch a part of their everyday lives. In the process, they naturally align and synthesize the functions of their chakras within the perspective of the higher orders of self. These students become significantly more intuitive, altruistic, and articulate. Their thinking becomes noticeably more focused and coherent. Sensitivity to others deepens, along with personal psychic sensitivity.

Today, when the stigma has been removed from talking with one's plants, it will not seem too extraordinary to learn that many who undergo these changes in awareness feel that they can also communicate with and understand other sentient beings, such as trees, birds, and animals, as well as they communicate with other human beings. Of course, these are the people who have a "green thumb" and whom creatures seek out. The story I am about to tell will, I think, serve to clarify this ability.

Communicating with the Intelligence of Trees

Trees, as I've mentioned, were my first friends. As I got older, I felt that I could have meaningful exchanges of thought with them using what I then considered different parts of my body, such as my throat. It wasn't until I became deeply involved in the intentional use of chakras during healing sessions, however, that I began to appreciate the specificity of consciousness of the different chakras, and to realize that there is a direct relationship between the particular species of the trees I communicated with and the chakras that were used.

Before that time, I thought my notions about trees were probably exaggerated and fanciful. A few years ago, however, I had the distinct joy and privilege of spending some time with one of the most powerful and well-known witch doctors in Africa, also well known for his artistic and literary abilities, Credo Vusa'Mazulu Mutwa.[18] In discussion, we found that we had several experiences in common regarding both healing and the teaching of healing. We discussed the development in persons studying healing of high sensitivity to all life forms, and the correlation between that sensitivity and the chakras one uses in communication with specific tree species. One of my students taped the conversation, which took place in Soweto, a native village outside of Johannesburg, and below is an excerpt of the transcription. We had been speaking about the effect human emotions have on plants:

D: What I would like very much to ask, sir, is: Do you yourself communicate with plants and trees? Do you find that you are able to communicate with different plants and trees with different parts of your body? For instance, I find that with trees I live with in the northeastern United States, in what the Native Americans call the woodlands, where most of the trees are hardwood trees, when I communicate with them I find that I do so with this part of me, just beyond or outside the throat.

C: Ah-ha! Yes, professor.

D: I am very anxious to speak to you of this, sir. In the far West, we have several very large trees that are many, many years old, the Sequoias.

C: Yes, the Sequoias; I've been there.

D: Yes, in the Muir Woods in the Bay Area of California, for instance. To my surprise, when I am there and I sit and meditate and I try to get in touch with the sequoias, then I find that I communicate from up here,

from the top of my head, from what the Far East Indians call the sahasrara or brahmarandhra chakra.
C: (Laughter.)
D: And when I am in the desert and I try to get in touch with the tall cactus—have you been to our deserts in the southwest?
C: I was in the Navaho desert.
D: Ah, yes! I know that area well. . . . but to go on, when I attempt to communicate with the tall cactus, then, again to my surprise, I find that I communicate through here, in the solar plexus chakra . . . and obviously from the look on your face, you do, too. (Laughter). This is what I would like to share with you.
C: You see, professor, let us say, now I am talking with one of those trees which grow near here (pointing to a group of trees). I find there is a slight cold in my feet, around here in my legs (pointing to certain secondary chakras in the legs). The more deeply we communicate, the more the slight cold rises. The same, exactly the same as you feel. This tree is not indigenous to Africa. We use this (other) tree for the treatment of rheumatism. Now, when I am trying to talk to that one, its spirit comes not to me through the body, but through the top of the head. Now, I thought I was the only one who felt this, but now I see others do, too.

As a result of that meeting, Credo had a necklace made for me, which he blessed over fire, and he also gave me a Zulu name, Uyezwa. This translates, I was told, as She Who Understands. I look upon this name as a kind of credentializing. The memory of this incident stirs up within me the realization that my experiences are actually real, and have a validity to which others can attest.

From the doorway of the sunroom in which I am writing, millions of acres of wilderness extend, ignoring the convention of virtual separation between Canada and the United States. (This area is now called the International Peace Park, so named because mutual backcountry has no formal barriers). These "other nations" of flora and fauna I count among my good friends. Several have distinct characteristics that single them out as individuals. Here in Montana, at the edge of the wilderness I mentioned above, often when I return from travels that have carried me to different lands, I feel a distinct shift in consciousness and am aware of a sense of greeting and welcome as I drive into the mountain from the airport. It lifts my spirits and lends a special, personal sense of acceptance to my homecoming.

Trees as High Prana Partners in Healing

Trees have an overabundance of prana that they are very willing to share, as anyone who is stressed may find, if they place themselves close to a tree and simply relax, for the tree's natural outflow of prana will do the rest. I found this out by chance when I was in the home of a young woman dying of cancer. She was utterly alone in her travail and had asked me to do Therapeutic Touch with her. An assessment of her vital-energy field made only too clear the direness of her situation.

My first instinct was to look around for someone to help me do Therapeutic Touch adequately for this lady. Other than the two of us, there was nobody else in the room, but my eye caught the sight of a healthy, mature balsam tree outside the open window. There seemed to be a moment of intense communion between us that conveyed to me the sense that the balsam would be willing to help. I linked up with the balsam's energy and directed it and my own vital-energy toward the lady without mentioning the tree to her. Whether this collaboration was an actuality or simply a symbol in my mind, the TT session went well, and I left the healee sleeping peacefully.

During the next several times that I came and we did a TT session, the healee's affect was much lighter, the quality of her life was more affirmative, and she even displayed a delightful sense of humor. One day she had enough energy to attend a party for a little while, and she told me, "I talked more there than I have to anyone except my new friend, that tree," nodding toward the window. I listened with astonishment, because I had never told her about my own engagement with the balsam. The lady died some weeks later from complications due to an overdose of radiation therapy that was mistakenly given to her. However, since then I have never hesitated to request the help of nearby trees when I do Therapeutic Touch, particularly when I do the session outdoors.

This story might seem strange to you, expecially if I told you that I've learned quite a bit about pranic flow from tree friends. There never has been an untoward result from these interactions, and so I recommend that you make friends with the next tree of your choice. They can be quite intelligent—after all, their bodies know how to capture sunlight and turn it into food, though ours do not. However, if you talk with them, don't expect them to speak English! They speak Tree, and you have to learn their language for it to work.

Force X: Nonlocality, the "N" Dimension

Transmitting Information over Distance

Nonlocality is a term ascribed to a timeless dimension where things happen "out of the blue"; that is, events seem to occur without cause, at least as we understand "cause"—usually as a machine-driven, push-pull process that mechanically or logically produces an effect. The concept of nonlocality arises from an assumption that consciousness has no boundaries in space or time. The idea has some generally accepted basis in the modern world, for scientists have been aware for several centuries that creative ideas and breakthrough hypotheses can occur at the same time to several people who are in different parts of the world.[19] From evidence such as this, the assumption has developed that the mind is able to transmit information over distance and to act upon it (at that distant place) under appropriate conditions.

There has been much anecdotal material about these notions, but now serious research is also in progress to explore them. The areas that give themselves best to the control demanded by current research methods have primarily been intercessory prayer and belief systems—their effect on health, and their powers of healing at a distance. These are serious, in-depth studies in valid research settings, often taking place in university medical centers. Other areas of current interest include healing at a distance, remote viewing, and the role of personal myth in the stories we tell ourselves to awaken our spiritual potentialities and call them forth. These last three will be discussed as representatives of Force X that may occur during the healing session. Of particular interest are the accounts of Mouse 37 and the previously mentioned study of Vivid Visualization. Since the time of these studies, several retests in various parts of the world have lent solid support to their validity and their reliability.

Healing Mouse 37 at a Distance

One cannot intervene in another's life, as one does when the charge is accepted to help or to heal, without convincing oneself that what is being done in the name of therapy is truly based on objective reality. Wishful thinking, impulse, fantasy, and exaggeration—which I have termed the Four Dragons (see p. 200)—have little place in the healer's consciousness if valid healing is to occur. Nevertheless, one can succumb to their urgings easily and sometimes quite innocently.

The control offered by contemporary research is considered by society to be one way to assure that an experience has some measure of validity and reliability. One of the healing experiences that lends itself to a wide range of personal interpretation is healing at a distance, in which the healer and the healee are geographically removed from one another. The healer is not at the scene where the healing action is taking place, and can create in her mind any scenario. The situation, therefore, can be highly vulnerable to misinterpretation, illusion, or even magical thinking.

Traveling about a good deal as I do, I often attempt healing at a distance for persons who are ill. From years of experience with meditation of various kinds, I have gained confidence in my own abilities. Nevertheless, the responsibility of teaching others what one knows stimulates a need for objective knowledge about the content. The Vivid Visualization process seemed to present an opportunity for such objectivity—but it all started when I got to know, at a distance, a little mouse, Mouse 37.

Several years ago the Nutrition Institute of America, under the directorship of Gary Null, Ph.D., conducted the largest single experiment to that date on two modes of healing—healing by direct contact and healing at a distance. Dr. Null phoned me to find out if I would participate in the experiment. Since the experiment began during the rush of the final weeks of the winter semester, and as faculty I could not take the time to go to the study site, I limited my involvement to the aspect of the experiment that was concerned with healing at a distance.

By prior arrangement with Dr. Null, I received a photograph of Mouse 37, one of two mice assigned to me. The second mouse was not given a number, although the accompanying note said it was Mouse 37's cagemate, and I assumed that it was a control. A brief letter of instructions asked me to keep a record of what I did. I was told that 174 genetically standardized mice had been injected with a highly virulent cancer virus and that the mice were not expected to live for more than three weeks. A total of fifty healers comprised the sample.

Since I ask my students to keep journals, it occurred to me to do the same. I drew up a list of what I thought might be useful data: date of each entry, the time, the context of the session, my daily activities, what I visualized during the session, the content of the meditation that I did as part of each healing-at-a-distance session, and other comments that seemed relevant. I continued to keep the journal for twenty-two days, although it was time-consuming. However, since I had no further communications from anyone connected with the research and was feeling the press of other commitments, I finally gave up the journal—but I kept up the healing at a distance for Mouse 37. My impression was that his cagemate had died several days into the experiment, but that Mouse 37 was well and hardy.

At the end of June I met a man at a cocktail party in San Francisco who was connected with the group doing the Nutrition Institute experiment. He told me that Mouse 37 was one of three mice still alive and that Mouse 37 and one of the other mice had no signs of cancer. I, of course, was delighted with the news. I had been doing healing at a distance for my little friend, whom I felt I knew quite well by now, and I continued to do so. In September—strangely, at another party in San Francisco (I did not live in San Francisco, and the study was being done in New York City)—I again received news about Mouse 37. He was well and very alive. Both of the other mice, however, had died.

Later, a year after the initiation of the experiment, I heard from Dr. Null that Mouse 37 was still hale and hearty. (In celebration, one of my friends offered to retire Mouse 37 to her ranch in Sonoma County!) This healthy state continued for several months until the following June. In response to a letter from one of my students, who planned to do biochemical studies on the effects of Therapeutic Touch, Dr. Null wrote: "Mouse 37 recently died of old age. It had lived longer than any of the other mice, and lived without discomfort and in a seemingly normal state of health."

The study gave me a deep sense of satisfaction. In that year and a half, my interaction with Mouse 37 had become quite real to me, and I had learned a great deal from the experience. The really interesting part of that adventure was the later verification of several visualizations that I had noted in my journal. One visualization I had was so clear that I could write a description of the laboratory in which Mouse 37 lived his days. I had occasion to go to the laboratory for the first time some months after Mouse 37 died, and Dr. Null took me on a tour. The number of flights of stairs up to Mouse 37's room, the way the litter cages were arranged, the color of the paint on the walls, the spatial relationship of the door to the windows in the room, and the direction of the building itself all corresponded to my previously written impressions. Moreover, I seemed to have sensed several of the healers who also were engaged in that study; that is, I had intimations of their presence or thoughts in the laboratory in the early days of the study while I was in the process of doing healing-at-a-distance for Mouse 37. Later, I appeared on various television programs and panels with several of these healers in different parts of the country. These occasions gave me an opportunity to check out my impressions with them directly. The verification of visual impressions seemed to me to be the most significant aspect of my own involvement in this experiment, and I was able to use this experience in developing a study designed to test the validity and reliability of such visualizations at a distance, which I called Vivid Visualizations.

The Non-Local Nature of Vivid Visualization

Visionary experience is ancient and is reported in all cultures of the world. For instance, a unique feature of the Native American culture is a strong dependence upon individual visions as basic guides in life decisions. In yoga, visualizations of persons or events at a distance are regarded as naturally occurring as the result of yogic experience.[20] Visualization is a subjective process, however, that continues to be poorly understood. Just how we are able to create pictures in the mind's eye so that the mind later recognizes them, whether those items are objectively present or are an abstract symbol, eludes precise scientific explanation. This is a paradox, for scientific methodology itself relies upon visualization as an introspective act as, for example, in the capacity to "see" a problem, the intuitive skill to clearly state a hypothesis, the capacity to perceive an analysis of data, and the genius to "foresee" the inferences of research findings—all depend on the capacity for visualization.

Although I have already discussed some aspects of the study in which nurses did Vivid Visualization (in reference to intentionality as one of its aspects, see p. 122), a second aspect of this study relative to the therapeutic use of nonlocality in healing at a distance is presented below.

This study on Vivid Visualization was the first to question whether the visualization of information about events occurring at a distance could be used with reliability in a therapeutic setting. In designing the study on Vivid Visualization we took care to learn from the experiences of Russell Targ and Harold Puthoff, physicists at Stanford Research Institute[21] whose work on remote viewing had provoked some controversy.

This study was designed to test the reliability of Vivid Visualizations that nurses (nurse-healers) had about hospitalized patients while these nurses were engaged in a meditative act of healing at a distance at a place geographically removed from the patients. The control group of nurses came from similar experiential and educational backgrounds. However, instead of having vivid visualizations while doing healing at a distance, they imagined the conditions, surroundings, and interactions of comparable patients who were hospitalized at remote locations.

Since I've mentioned this study in the section on Force A and the study itself is in the literature,[22] I will not repeat all the details of the protocol, but will simply reiterate that the task assigned to the experimental group was to visualize themselves at the bedside of the patient, treating him or her with Therapeutic Touch. While they were "there" they were to note several items about the room and about the patient, and after the healing-at-a-distance session they reported on these items and their experience on standardized forms. Each nurse had two patients to report on, and the study went on for three consecutive days. For both the experimental and the control group there were other nurses on the scene at the time of the sessions on each of the three days, who acted as observers and also reported on the patient's interactions with others, the equipment in the room, and any changes that occurred over the three days in the patient's condition, features of the room itself, or the equipment in the room.

The level of statistical significance for testing the hypothesis of this study had been set at $P > .05$ (i.e., the possibility that the results of this study would be due to chance was judged to be less than five times out of one hundred trials). This decision was based on the sample size and the recognized complexity of predicting behavior of even a small facet of human consciousness. The actual level of confidence for all hypotheses far exceeded $P > .05$ in support of the hypotheses. This robust level of confidence has served to encourage continued research.

From this study, it appears that nurse-healers who have a deep-felt concern for their healees, whether in their presence or at a distance from them, would be prime candidates for the enactment of Vivid Visualization. However, it also is understood that nurses are not the only people who deeply care about others. Compassion is, of course, a generalized human potential, as is Therapeutic Touch. Therefore, it seems clear that all persons who actualize that potential for compassionate healing and concern also have a high potential for Vivid Visualization.

A Random Example of Vivid Visualization as Process

Several qualitative aspects of this study offer insight into the process of Vivid Visualization, particularly regarding the clarity of perception at a distance, the verification of which was the main thrust of the investigation. For this purpose one report is selected at random.

In this study, the only requirement of the person playing the role of nurse-healer was that she have had some experience with Therapeutic Touch and meditation. G, who was the nurse-healer in this study, and R, who was the observer, were a strange team. G was supervisor of an in-service teaching unit at the hospital where she was employed, and R, a male nurse-healer, worked in a coronary care unit of a hospital many miles away. R, the class skeptic, was reserved, cautious, and pragmatic in his view of the world. G was in the throes of great changes in both her lifestyle and her worldview, but she had not fully integrated these perspectives of life and living. The two seemed to have little in common and did not know each other before the study. The day G and R reported on their experiences was the first time they had had an extended conversation. "We had spoken fifteen words to each other in our whole lifetimes," said R.

G's Vivid Visualizations turned out to be remarkably accurate, although she said while doing it, "I felt like a fool" and felt threatened by the situation "because I didn't really know if my perceptions were right, and yet I felt I had to go on." Her method of doing Vivid Visualization was "to attempt to put (e.g., to feel) myself in the room with the client after centering myself" preparatory to doing TT at a distance.

G and R did not communicate with each other during the three days of the study. After the completion of that time, both of them met with me to read their reports of their experiences with Vivid Visualization into a tape recorder. It was then that they learned of each other's experiences for the first time.

As noted, both patients G worked on were under R's observation in the coronary care unit in which he worked. The case we shall look at was unusual in several respects. In the protocol, each nurse-healer was given two pieces of information about each of the patients he or she worked on: the name and the medical diagnosis. However, because of the rapid turnover in the CCU, R had to choose the patients on the day they began the study. Nevertheless, G identified the patients correctly. In the case we shall look at, G correctly described the patient as a frail elderly gentleman with blue eyes, white hair, and a full beard. Although he appeared helpless in her visualization, he radiated a strong sense of calmness and serenity. He seemed very intelligent to her, and she could not understand why she sensed that nobody talked to him. (This was because he spoke only Yiddish, which nobody on that unit could understand.)

G visualized that he was the only patient in a small room and was in a bed that was not near a window. There was a second bed in the room, she reported, which was empty. (The patient had gone to the Recovery Room that morning). G correctly reported that there was an intravenous tubing inserted in his arm, but other equipment at his bedside was standard and not at all unusual. However, G noted considerable changes in the bedside equipment the next day, and correctly identified what they were. (The patient had suffered severe pulmonary problems during the night and now had a tracheotomy with supportive ventilatory and suctioning equipment.)

G and R had agreed to participate in the study at 10 a.m. for three consecutive days. However, on the third day G had a prolonged staff meeting and then forgot about the study arrangement until she was on her way to lunch.

When she remembered, her first reaction was casual; it was too bad to have missed the session, but nothing could be done about it now. Nevertheless, it continued to dog her thoughts as she walked toward the cafeteria. Finally, much to her own surprise, she left her companions and went back to her office to carry out the meditative healing at a distance, even though R would not be observing, she thought.

Strangely, in the meantime, at the hospital several miles away, the coronary care unit on which R worked had become very busy that morning with an unusual load of emergency procedures. R's attention was caught up in the immediacy of the emergency situations, which did not let up until lunchtime. On the way to the cafeteria he realized that he had forgotten to do the observations that morning. However, the traumatic nature of the events he had experienced earlier was such that he dismissed the project for the moment as being trivial. He found himself thinking, "Oh, the hell with it. The whole thing is silly. I'll just fake it."

Nevertheless, after a moment's reflection, he turned on his heel and went back to the CCU to observe his patients. In this way, although neither G nor R knew it at the time, R was observing the patient at the same time as G was doing the healing meditation.

G visualized a young man stopping at the doorway and talking to the elderly gentleman. She correctly stated in her report that the young man had once occupied the other, now empty, bed in the room. (He had been transferred to another area of the hospital and now was being discharged. He had stopped at the unit to say goodbye.) G was perplexed because it seemed that the elderly man was obsessed with the idea of obtaining a straw. As G reported this when the three of us met to audiotape their reports, there was a highly audible gasp from R. It turned out, he told me when he could catch his breath, that the medical doctor had told the elderly gentleman that he would remove the tracheotomy if he could learn to drink through a straw. There were no straws in the CCU at the time and R, returning to the unit just then, was recruited to go down to the Central Supply Room to get the patient—who was making a tremendous fuss—a supply of straws.

In her journal G had written:

"On Thursday evening I volunteered to be part of a project I did not fully understand. I said I knew how to meditate, but had I really been doing it? Not recently, that's for sure. In my notes I documented feeling ridiculous in my attempt to visualize clients I had never seen. I felt that my visualizations were all contrivances of a vivid imagination. Yet, there was a strange feeling of peace and oneness with the clients and with the male nurse, R, who cared for them.

"When we reported our experiences finally, there was nobody in that room more amazed than I at the results. I certainly did not feel I had cured anyone, but I did feel a surge of confidence and a sense of validity that I had not identified with ever before. My faith in my own potential was not restored, I would say, but rather that at this late date in my life it was truly initiated."

R's comment when he heard G's report was, "I am a little flabbergasted. Her descriptions are extremely accurate. How could that be?" In his journal he wrote:

"G and I had never spoken to each other before the Thursday evening in which we took ten minutes to agree on the schedule of observations. Apart from her knowing that I worked in a cardiac unit, she knew nothing about the patients I would be observing. At that time neither did I know this, since I had no idea which patients would still be on the unit the following Tuesday, the day we had selected to begin the study. Our only other contact was on the day we gave our final report to Dr. Krieger.

"I was utterly astounded at the accuracy of G's description of the patients I had observed. I have never been skeptical of the assumptions that underlie the premise of this study. Rather I felt that the real question touched on which of the experiences described in the report could bear the weight of careful investigation and documentation....

"Sitting there listening to G paint an almost perfect picture of the patients, including their clothes, ages, personalities, lengths of stay and even snatches of things they said, placed me in the uncomfortable position of one who is at the heart of a profoundly moving experience, but wonders if those whom he will tell of it will believe what he is about to say. Douglas Boyd[23] has captured my sentiments exactly:

'Had I been listening with all my intellectual analytic habits, this feeling would not have taken hold. I would have heard it too literally and gotten caught up in comparisons. But my mind was off-guard.'

"Those last six words may be the key that will open my whole being to the 'mysteries' present here. Truth, and those aspects of reality that we are not able to characterize, nevertheless undergo trauma when we listen too analytically, too intellectually—not because such reality cannot stand close scrutiny, but because our analytical faculties, our intellect itself cannot come to grips with the immensity of their being."

As noted above, this study has been supported now by many replications here in the U.S. and abroad. To me, this not only attests to the reliability of the study but also forces the recognition of how little we understand about the capabilities of human consciousness and the potentialities of the therapeutic uses of nonlocality.

Personal Mythology: The Stories We Tell Ourselves

"The images of myth are reflections of the spiritual potentialities in every one of us. Through contemplating these, we evoke their powers in our lives."—Joseph Campbell[24]

A light snow is falling in the gray half-light as we gather around the open fire pit. The fire pit is newly constructed, and its most striking feature is twelve equal-sized round stones, plus another at the opening of the fire pit to commemorate the thirteen full moons of the year. We are an extended family of about forty people intent on the Earth Renewing Celebration in the way of the Apache, by whom I was adopted several years ago.

The Firemaker approaches and builds a sacred fire, calling upon the elemental spirits of the Four Directions and the magical Points in Between to feed the Fire so that its warmth and light will reach out to the People.

My sister, Oh Shinnah, who is conducting this ceremony, steps to the head of the Fire, faces the east, and gives voice to a dedication of this ceremony to Earth Mother. She then kneels down and draws a square-headed female Kachina, an ancestral spirit, in the earth. The Kachina's elbows are bent and her arms reach toward the earth to help bring down the energy for the People on this Day of Least Giving. Finally, my sister draws a heartline from the Kachina to the Fire. She then rises, nods to me, and points the eagle feather she holds in her hand toward the earth, just above the head of the Katchina.

I circle the Fire in a clockwise direction until I come to the Kachina. I portray the Grandmother in this creation story, and I kneel down just above the Kachina's head and dig with a large crystal through the frozen snow to the soil beneath. As always, I am surprised by the relatively warm, sweet odor of earth as it is exposed to the air. As I dig, I tell a story to the gathered People.

This is Earth-Renewing Day, I remind them, the time of the winter solstice. It is the signal for Father Sun to begin his trek northward across the sky to once again renew the life force in the seedlings of the Earth so that the People and all creatures might live. All people know of this ancient covenant, told to them by the People Who

Came Before, and as I dig I describe how, beginning in the easternmost reaches of Earth, as the world turns and Father Sun rides over the horizon, people all over the planet wait expectantly for the first slant of Sun's life-giving rays.

As the angle of declination widens and Father Sun's light sweeps the darkened Earth below to illuminate all it touches with its radiance, each gathering of people voices an expression of gratitude in their own language, each translating from out of their own belief system, each with the same message: The Sun returns! We are blessed once more with life-energy. It is like multiple waves of thanksgiving flowing over the Earth in a natural order of succession; a great, successive orchestration of people of like mind that spreads its jubilant sounds in ever-widening swells of joy, hope, and wonder in happy attunement with and in celebration of the first stirrings of the rebirth of all Nature.

I affirm to the People that that song of elation is our song. At this moment we are at the leading edge of a great rolling swell of jubilance and gratitude, and I encourage everyone to send out that rising personal sense of joy to join the exhilarating wake of Father Sun's travels across the face of this planet. We are at one with the rhythm of that spreading wave of exultation; its pulsations drive our lives to new beginnings and fresh understandings.

I place the six grains of multicolored corn that I hold in my hand—yellow, white, blue, and red—into the hole I have made, and then step back into the circle as the rest of the story is carried on by others. I turn toward the rising sun, and as its rays strike my face I suddenly realize that I wasn't just telling a story. I feel the power of the accumulated paeans to Father Sun as the pulsations, building with each successive voicing of gratitude, sweep over the Earth to the edge of our circle. Then, further empowered, the Sun journeys on to complete his mission.

In that moment I am so identified with the continuing vibration of wave upon wave of glad tidings that are traveling westward over the face of Earth, that I hear the voices of those telling the rest of the story resound with a soft echo, as if we were in a natural tunnel. As each speaks her or his part, it is like the reverberating voice of a huge drum telling this never-ending story of recurrent rebirth to the age-old mountain, and I am the story.

The Healer's Journey: Healing Oneself through Personal Myth

It was a brilliant morning as we drove through the foothills of the Berkshire Mountains on our way to Pumpkin Hollow, where my friend Emily was going to give a lecture. Emily had a mind that sparkled with brilliant and often original thought, and a heart that shared with everyone. As the car tooled along at a leisurely speed, I had a memorable time engaging her in conversation. We talked about the topic of her lecture: how to go about our daily life while listening to the inner promptings that well up from deep within ourselves.

From notes I scribbled down as we got into our discussion, I see that the direction we were going was that in the final analysis, as individuals, we never really see how our life ties into the whole, how the universe contrives to bring us forth into this mundane world and how we fit into the larger picture in a meaningful way. What, we pondered, are the core reasons for our actions and the manner in which we act? And what do our daily acts in the world disclose about our inner self?

The impact of that penetrating discussion has never left me, and since that time I often have had cause to wonder: Just what was the universe's purpose in bringing me into life at this specific time and in this particular place? And, as do all persons who have considered that question, I further ask myself: How can I fulfill that purpose? Involuntarily, a shiver goes up my spine, and I realize that another question is still lurking in my brain: Do I really want to comply with that scenario? Can I really live up to that purpose and give it meaning?

Each of us is the heroine of her own story, a story that is urged on by an insistent quest for meaning—for if your life has meaning, it has purpose and harbors the gift of intentionality. In this search, it becomes important to acknowledge who you really are, and strive toward understanding the archetypal forces that are modeling your behavior, the fundamental ideas that empower your reason for Being.

Archetypes teach us how to become full human beings. Even as this new millennium opens before us, our Age is forging new archetypes out of the raw, frontier stuff of our Becoming. There are hints in our daily acts of living that inform us of the archetypes to whom we bind our allegiance. For instance: On weekends and holidays, in the time you have not committed to a socially required lifestyle, who do you want to be? In the quiet of an evening at home, what are your favorite stories, TV programs, daydreams, reveries? What is your favorite memory? Your worst nightmare? Your ambition? Who would you like to be your companion on this Journey—the person to whom you could talk without inhibitions? In whom you would be willing to confide your real reason for wanting to help, to heal?

The good books say that the Journey begins with a call to leave a stage of personal growth that we have grown beyond or that no longer serves us well. The next stage, paradoxically, is a refusal of that call; we put off the journey because it seems to come at an inconvenient time, or because we don't feel up to it. The Warrior-Healer, however—an archetype that we'll discuss in more detail in the next chapter—finally heeds the call and embarks on a journey of self-exploration, leaving the realm of ordinary reality and turning her attention to inner visions and the more interior life.

Usually, however, a force that is often unfriendly keeps the seeker from delving deeply into the world of non-ordinary reality. This force frequently comes in the guise of firm schedules, fixed habits, and rigid attitudes and may be personified by authoritative figures, e.g., supervisors, or demanding relatives or friends who insist that relationships remain as they have been and fight your attempts to change. To escape their insistent pull back toward outgrown ways, the questor needs to be cunning, clever, and committed to her inner Journey. In time, however, it is realized that the real imprisoner who refuses entrance to other realities is oneself. It requires all one's wits to overcome self, the personality, which is so often caught in reflexive behavior, and finally cross the threshold to the inner world.

Once able to conquer this resistance and enter a new realm, one recognizes that previous identities must be altered or given up as a new life is acquired. This realization can throw one into a time of grieving for things and events now receding inexorably and irretrievably into the past. This may be cause for sadness and depression or darkness of spirit: we may want to run away, forget the Journey, and just be as we once were. This is a time of unusual tests, many trials of spirit, and incredible challenges, but somehow one finds the inner resources to survive and resume the Quest.

It is at this time of seeming darkness and tribulation that we may find an inner source of unusual helpers and allies that slip through from another reality to help or to heal us. It is as these potentialities are put into service in our lives and become actualized that we begin to stride forward with confidence on the road to personal transformation.

Heartened by the promise of attaining an inner state we have yearned for, we step confidently on the path in quest of a greater and more profound expansion of consciousness. It is this personal knowing that will bestow a deeper grasp of what it means to be healed. Moreover, there can be a concomitant understanding that our newly-recognized liaison with the inner self may be used in the service of helping or healing others who also yearn for this inner reality that is so difficult to find, but is always present, "nearer than hands and feet," closer than the heartbeat, and so obvious in retrospect—how did we not know?

SUMMARY

In these chapters on little-noted subtle forces that significantly affect the healer-healee therapeutic interaction, we looked at healing as a humanization of energy, particularly as an interior expression of human-energy force fields as they are available during the TT session. The human energy field is uniquely suited as a medium for healing self and others because of its innate abilities to organize itself and to replenish spent energies, and because its energy levels interpenetrate each other. This last factor means that during the TT session, what is done to one level of consciousness naturally and spontaneously affects the other levels as well, thereby leading the healing process toward a unified effect on the healee.

Prana in its many aspects is the ground or foundation for all of these levels of activity. In the physical body, prana becomes physically functional as it arises into materiality at the core of each cell; that is, in the nucleus, probably at the site of the mitochondria, the bodies responsible for cellular respiration and bioenergy production. Since prana is nonphysical, it can leap or interpenetrate the cell membrane wall and thus bring its lifegiving vital-energies to the organism as a whole. The rhythm, reactivity, and constant movement of prana within the individual create a unique vibratory signature for each person that continues throughout her or his life.

In the healer, there is a decided acceleration of affect, an increase in sensitivity, at the psychodynamic level of consciousness as the field becomes more pliable and supple, responsive, and sensitive to subtle energy patterning. The healer's conceptual field also undergoes changes of a transcendent nature as her commitment to come into conscious engagement with the inner self becomes grounded in intentionality. Intuitive insights, meaningful visualizations, and creativity may then become an integral part of the healer's resources to help those in need.

Finally, we begin to see that several forces are at work during the Therapeutic Touch healing session (here simply named Forces A through D and Force X), and we begin to catch a glimpse of how we organize and coordinate these conscious and unconscious experiences. Force A represents the energies generated by the higher orders of self during the TT interaction. Force B derives from the state of consciousness of the healee and his therapeutic interaction with the TT therapist. Force C is produced by the sacred or committed place of the healing site and the power inherent in it. Force D represents the direct helping intervention of natural powers, and Force X includes non-local effects, such as acts of Vivid Visualization, and the stories we tell ourselves.

In using these forces purposefully, for a wink of time we have an opportunity to reach for, and perhaps touch, the interface with the nonphysical implicate order that makes healing possible, and to know with certainty that these forces are human potentials we can actualize to bring their creative energies to the healing moment.

PART III

THE TEACHING/LEARNING FRAME OF REFERENCE: TAPPING THE CORE OF THERAPEUTIC TOUCH

CHAPTER XII: THE WARRIOR-HEALER

The Warrior-Healer in Modern Dress

My mind's eye travels to the hospice, where so many TT therapists have found deep satisfaction in practicing TT to ease final transition. This setting has encouraged the development of several thoughtful, innovative, and life-affirming but death-accepting healing techniques and relational behaviors. This may surprise the uninvolved observer, for more often than not the first view the TT therapist has of the hospice patient in the throes of final transition is of a knot of contorted, emaciated body mass entangled in a johnnie or janie coat much too big for the shrunken body. The person is exhausted from the encroaching disease process and the iatrogenic effects of chemotherapy, radiation, and other heroic medical regimes. The sour odor of vomitus or the sounds of retching cling to the surrounding atmosphere, and sometimes there is evidence of urinary incontinence. There, one realizes, lies the most dismal embodiment of forlorn hope and vain expectations. The patient may be overwhelmed by a sense of hopelessness and helplessness. Deep sadness and melancholia often are reinforced by devastating feelings of aloneness and utter emptiness.

Why would someone in the depths of such dark despair request the aid of a TT therapist? And how does the TT therapist continually renew the courage to return? How can she participate in this pain and fear; where does she find the power of soul? She derives her strength from the act of deep, sustained centering, that ability each of us harbors to commune with our own depths for the insight to help those in need. It is the power of her own spirit from which flows that inner strength to compassionately help others and yet not falter oneself. It is a true expression of the Warrior-Healer.

Warrior-Healer? Is the very thought a paradox? In all major cultures on this planet, the concept of Warrior as archetype is as common as that of the Healer, and there are designated rites of passage for each. Contrary to most opinion, the way of the Warrior has nothing to do with violence, or the desire to destroy, or the imposition of one's will upon others. The war, if that term is even applicable, is always against one's own weaknesses and self-imposed limitations, and against those in society who would throttle others' rights to similarly grow in the image of their innermost selves. The Warrior is adamantly committed to this self-search and unyielding in the pursuit of gaining control of her personal world.

The Warrior works toward her goals by consciously taking into her own hands the direction of her personal life. She does not rush impulsively into life, but learns to carefully balance inner and outer reality. The Warrior will do whatever is necessary for the good of the whole, and whatever is necessary to bring about harmony and peace as a result of her actions, but she will not compromise. In the process she learns the lessons of humility, respect for the orderliness of the universe, and appreciation for the dignity of compassionate regard for those in need. In every sense the Warrior is worthy of being a conjunct of the Healer. The Warrior-Healer is the integration of two sides of the same coin, a full expression of the commitment to the conscious workings outward of the mission of the inner self.

Healing vs. Curing

The Warrior-Healer recognizes the essential nature of her mission by acknowledging the real distinction between healing and curing. The words even have different derivations. The term "cure" comes from the Latin *cura*, meaning to remedy or get rid of an ailment or evil. It refers to a course of medical treatment, a panacea.

"Heal," however, comes from the Middle English word *haelen*, to make whole, and is further defined as "to make healthy, to bring into harmony." Healing has to do with how one thinks of oneself, one's worldview or philosophy of life; it implies an enhancement of the quality of one's life. Fundamentally, to cure is to care for, but to heal means to care about, which implies liking or affection.

From the point of view of the TT therapist, healing entails a conscious, effective engagement of the self in the compassionate interest of helping and being fully present to one who is ill or in trauma. It is within this context that the TT therapist gains the courage for an attitude of death-acceptance for those in final transition while simultaneously projecting a frame of mind that will support a recuperating person with life-affirming verve, vigor, and vitality. She acts appropriately in either situation, ever cognizant that an outcome of life or death is not a measure of her success or failure; rather that the consequences of illness are exceedingly complex and that we are largely unaware of their root causes.

The Transition: Warrior-Healer to Warrior-Teacher

It is these unknowns that make the TT interaction a tough discipline for the committed therapist. The person who wishes to make the practice of TT a way of life has to be willing to explore senses other than the six usual ones, which is to say that the TT therapist must be open to other realities than that accepted by the dominant society. It is the compelling shift in consciousness that closely follows the act of centering that immerses the therapist in this other reality, in that other state of consciousness in which, driven by compassionate intentionality, she can gain direct access to the healee's vital-energy field.

In the challenge to heed the realities at the core of the centered state while involved in the compassionate concerns of the healing moment, the practice offers the therapist a new consciousness in the service of those in need. The insight born of centering offers a rare opportunity to explore the fullness of one's own human stature. In that examination, one realizes radical sustaining kinship bonds to others and feels the need to share with them this healing lifestyle that offers them the opportunity for their own self-realization. Indeed, the desire to share one's experience becomes a metaneed, an urgent need to help those who lack the healing insight by teaching them this most humane of all human undertakings.

When one accepts the responsibility for cultivating the balance and self-control that are necessary prerequisites for that mindful state, one becomes a special kind of instrument. As she hands on this torchlight of learning, of healing, to kindle in others our common illumination and help them sustain that congruent vision, the therapist becomes a Warrior-Healer/Warrior-Teacher.

CHAPTER XIII: TO BE A WARRIOR-TEACHER

Teaching Therapeutic Touch

Therapeutic Touch sits in two worlds. One world is establishment-oriented, for the theoretical content of Therapeutic Touch has blossomed under the fostering touch of the university, the hospital, and the community. The other world of Therapeutic Touch is future-oriented. In one Ivy League anthropology department, the teaching style of Therapeutic Touch was put forward as the suggested way in which community health teams should be taught in the future. It has also been suggested that the TT assessment could foster the development of a new therapeutic language.

The beginning of Therapeutic Touch itself arises out of two universes of discourse. One aspect derives from the findings of basic research, theory development, and the world of literature on human consciousness, subtle human energies, and theories of illness and of high-level wellness. However, Therapeutic Touch also spans an experiential spectrum of human interaction that extends from basic physiological reactions to inspired responses that edge into the realm of the transpersonal. Experience on this level forces the recognition that living beings do not have hard edges, but that their human essence progressively fades into eddies of subtle energies. These vital energies flow in patterns whose sources are in n-dimensions and x-time, a timeless hyperspace nexus easier to experience than to describe.

The therapist herself must face, Janus-like, in two directions at the same time. Objectively, she confronts the questions of what the matter is and how the healee can be helped. However, the search for true intuition on the derivation of the problem and the significant relationships inherent in it is a subjective one. The answers to these questions are unique to each situation, for they rely upon the TT therapist's past experience and the extent of her inner work, and are shaped by a joining of her personal knowledge with the depth of her compassion for those in need.

To Be a Warrior-Teacher

To be a self-confident teacher, take as a basic assumption that the students who come to your classes want to learn. In order to fulfill their expectations of you, you must overcome your fears and apprehensions. Just as the TT therapist strives to be a Warrior-Healer, the teacher of Therapeutic Touch must aim to become a Warrior-Teacher. Set your own standards. Do more than is expected. Teach in a life-affirming, indeed, a life-celebratory manner. Be a model for your students. Strongly put forward the thought that in all ways our reach should exceed our grasp, and let them emulate you.

Dynamically engage your students in the learning process. In Therapeutic Touch, experiential learning—literally allowing the energy to "run through you"—is very important; emphasize the "work" in "workshop." Strive to give your students many opportunities to experience under your supervision the wide range of vital energies in living systems; this will allow them to sharpen and deepen their perceptive and evaluative faculties in preparation for the TT assessment. The true focus and greatest contribution of the teaching of Therapeutic Touch is the awakening to the therapeutic functions of the vital-energy field. Make it a personal "ah-ha!" experience for the student.

Extend the student's vision of him- or herself—and then sit back and note with ever recurring wonder the astonishing change in the class atmosphere once that personal breakthrough occurs the first time. "I do not stop at my skin!" the student will exclaim in an epiphanic moment of personal discovery, having sensed the previously unrealized depth of his or her being. It is this astute perception that fires up the fever of their learning and sparks the enlightening stream of your teachings.

The Mindless Dragons: A Word of Caution

Although you are encouraged to pique your students' interest and curiosity, stimulate creativity and innovative perspectives, and encourage intuitive insights, remember that mindless euphoria, deceptive illusion, and unbridled energy are enemies of Therapeutic Touch. It is in the interest of the validity of your teaching that you insist that students recognize the responsibility they take on when they become healers: they are intervening in another's life. Insist that the student remain grounded—at least eighty percent of the time!

I have often thought of the most potent of personal polluters—Wishful Thinking, Impulse, Fantasy, and Exaggeration—as the Four Dragons. Dragons project energies—fire and brimstone—that can be hurtful, and, as all myths affirm, the only one who can kill the dragon is the one who must confront the dragon, that is, oneself.

Recently, as I was thinking in this figurative manner, the metaphor expanded. I realized that the Four Dragons can have offspring, negating emotions that flow from these dragons when they are allowed to flourish: Wishful Thinking begets Anxieties, Impulse gives rise to Insecurities, Fantasy may give birth to Paranoia, and Exaggeration can issue forth a flock of Defenses.

These dragons and their offspring are equally damaging to the healing moment, for it is natural to project from our own vital-energy field emotional as well as physical energies. If these dragons arise during our healing interactions, their attendant emotions can accompany the TT therapist's projection of an intentionality that is meant to help the healee, hurting him instead.

What can be done to reduce the threat of these monsters of our own creation? Counteract Wishful Thinking and its consequent Anxieties with Intentionality: negate Impulse and its resultant Insecurities by Centering the Consciousness; deny Fantasies and uproot tendencies toward Paranoia by practicing Clear Visualization; and diminish the power of Exaggeration and its claims of impenetrable Defenses by instituting measures of Replicability, particularly when recounting anecdotal reports of healing or helping through Therapeutic Touch.

Table V. The 4 Dragons, Enemies of Therapeutic Touch*

THE 4 DRAGONS	THEIR OFFSPRING	RESOLUTION
Wishful Thinking	Anxieties	Intentionality
Impulse	Insecurities	Centering of the Consciousness
Fantasy	Paranoia	True Visualization
Exaggeration	Defenses	Replicability

*A mnemonic for each set may be helpful:

1. The four major dragons: WIFE
2. Their offspring: PAID
3. Resolution: TRIC

Responsibilities of the Teacher of Therapeutic Touch

Teaching should be a constantly new learning experience for the teacher as well as for the student. Accept the challenge by striving to extend and deepen your own knowledge every time you prepare lectures; if what you are saying bores you, it is sure to bore your student.

One way to keep alert is to question ideas from related fields for relevance to what you are teaching. If the material is appropriate, integrate it into your own presentation; however, be sure you understand the context of these ideas so that you don't misrepresent their initial intention.

Establish a personal reference library out of the wealth of books, videos, CD-ROMs, and other electronic teaching tools now available. Set up files in your computer of authoritative papers on healing and communicate with the authors on questions pertinent to their areas of expertise. Network widely when you are at conferences. Follow up people and data that may be important to your interest by email, fax, and other avenues of the electronic and digitized media. Keep a cross index of your teaching notes in your computer to facilitate easy referencing and exchange of ideas on the current information highways, and freely use bulletin boards, search engines, salons, and chat rooms to keep a little ahead of the traffic of day-to-day information flows.

Finally, even though Therapeutic Touch is a subtle touch, it is of course an intervention, and therefore we must accept personal responsibility for the consequences and the inferences of our people-to-people interactions. However, the beginning student, too, has responsibility, even if only "to intuitively fill in the gaps left by the teacher." To help in this, several examples of homework will be included in the last chapter.

Enhancing the Effectiveness of Your Teaching

As you begin to teach, review and critique your own presentations periodically to insure that you do not fall into a routinized, robotic teaching style. Frequently challenge yourself by sitting in on other teachers' classes and engaging in peer review with your colleagues who are giving similar courses in other localities. Stay up to date on current literature and research in the field; seek out new ideas and new problems or contemporary restatements of old ideas and old problems that have not been adequately addressed.

Do not teach simple factual material only; thoughtfully integrate concepts that may bind facts together, and seek out their underlying principles. Use relevant illustrations; if they are not germane to the point, have the courage to leave them out.

In class, be sensitive to how you are affecting your students' ability to understand what you are presenting. The ability to present material clearly, concisely, and correctly, and to change your mode of teaching to meet the needs of the students, is an art well worth the time and perseverence to develop. Keep in mind that the raison d'être of teaching is that the student clearly understand, and remember that students rapidly pick up on a teacher's anxiety and fear and may follow her example. It's best to be casual and unaffected. You know more than they do, but that may be only because you got there first.

Do not be in competition with your students; reward them when they are right, bright, and questioning. Ask yourself frequently, "What can I do to enhance my students' learning?" "What do I do to detract from that learning process?" "Are my classes really effective—do the students learn? Do students enjoy my classes?" How do you find out? Distribute brief but searching evaluation forms at critical points in the curriculum. When you read them, make notes of relevant comments and ideas for referral when preparing your lecture notes for future classes.

An important component of teaching effectiveness is transmitted through body postures and gestures, so be aware of your body language while teaching. Modulate the volume and monitor the inflection of your voice appropriately as you speak, and never mumble—give the student every chance to understand you. It will be difficult not to use gestures, since you will be talking about hand and hand chakra functions; however don't overdo it, for of all body language, messages expressed through the hands will make the strongest impression on your students. Hands-on demonstrations are very effective and will permit you to get out from behind the desk and use movement to demonstrate, supplement, or emphasize your teaching points. However, be aware of your energy level as you do so, and note the influences the teaching content has on the students, especially on their mindset regarding healing as a therapeutic interaction. Be aware of the constant slipping apart of intergenerational communication modes and the new meanings that may inhabit old words in these changing times.

Effectiveness in teaching requires confidence. As has been discussed, increasing confidence is one of the indicators of the TT therapist's ability to make an ally of the inner self, and of her understanding of the responsibility that relationship entails. However, if you want to display confidence while in front of a class, don't try to teach more than you yourself know and have experienced. Therapeutic Touch looks simple, but is quite complex as one proceeds deeply into the process, and experiential knowledge becomes a priceless guide.

Enthusiasm for the subject and personal warmth are also important to the effectiveness of your teaching, as is stability, which is marked by poise in one's manner. To maintain poise while teaching difficult topics, I often over-prepare for the class; then, once I'm sure I have it right, I state the material simply (but not simplistically), clearly, and casually. This ploy allows me to perceive what I am about to convey from a non-threatening perspective and to concentrate with confidence on helping the student to understand the material, too. In presenting material that may be abstract, I try to refer to various examples of relevant real-life situations. I try to keep the story short, particularly if it describes my own experiences. It is not a good communication tactic to describe in great detail personal experiences that may be of interest only to a few. To help in the acceptance of complicated matter, of course, visual aids are very helpful.

To check that I've kept the content on target and its organization acceptably logical, I often will tape my lectures on new material and play them back as I drive home, noting where I've gone astray. To check on my behavioral styles, I will have myself videotaped while I conduct a class, and watch the video with the sound off so that I can analyze my behavior more clearly.

Practice the intentional soft sell: throw away a line so that it looks easy and seems learnable. For instance: "The act of centering the consciousness is not difficult and it is enjoyable—because it is all about you!" To soft-sell, know your material thoroughly, and be a master of timing. This means not underestimating the depth to which TT can go as a human therapeutic interaction, but remembering to include touches of humor, an ancient teaching strategy and an indispensable ally. Humor allows the mind to shift away from stolid rigidity, to shake up the neurons binding an idée fixe, and to open to just one more thought. Like all the tactics discussed here, its objective is to engage the student in the learning process, for that is the whole point—and the joy—of teaching.

Reflections

To be a teacher, you must want with a passion to help those less knowledgeable than you to know what you have learned—a passing on of the torch of knowledge in the best traditions of civilization.

To be a teacher, you must find the courage to be an actor, so that you can project the material you are teaching as bona-fide reality—a seeming oxymoron.

To be a teacher, you must be able to enjoy what others may think of as the tiresome and sometimes tedious task of whittling concepts down to the level of the students' readiness, without dumbing them down or taking the concepts out of context.

To be a teacher, you must cultivate self-discipline as well as commitment—be willing to strive for excellence in yourself as well as in your students—for to be content with mediocrity is boring.

To be a teacher, you also must be willing to be a perennial student, always ready to learn more, to learn the new, to learn in depth, to learn the unexpected.

Finally, teaching requires a striving in two directions—toward understanding in relation to your own world perspective, and toward clear communication of your insights.

Summary

There is a strong relationship between healing and teaching. Under the best circumstances, the healing process can be educational (from *educere*, to draw out or evoke that which is latent or slumbering) and the learning process can be healing (wholing). Within this frame of reference one can perceive the healing act as an opportunity to reeducate the healee, to change his views or beliefs on how to maintain a state of health. The role of healer thus encompasses that of teacher.

When instructing students in the techniques of Therapeutic Touch, a process that will be explored further in the next chapter, it is important to instill in them the awareness that it is only by skillful practice, plus the critical analysis of the process and the willingness to learn from that investigation, that one is enabled to reach deeply into the consciousness, where the true TT process lives. That is the focus one should have in doing Therapeutic Touch: to become the centered, living embodiment of the healing moment in conscious transaction with one's inner self.

The Warrior-Healer/Warrior Teacher knows that each healing is a new experience. Each person's subtle energy fields at the various levels of consciousness—the combinations of wavelengths emitted by each person's living flows and patternings, radiating luminosity into the nonphysical media of his or her personal surround—are unique. And each time one person helps another to heal, the wonder of this miracle blazes forth, and we recognize that we haven't come to the end of our topic; we've barely tapped the roots of Therapeutic Touch.

CHAPTER XIV: AWARENESS OF LEARNING STYLES

Coordinating the Learning Process with the Healing Process

In ancient Greece, students—for instance, those attending Pythagoras' school at Krotona, the *akoustikos*—learned through oral lectures. Today research findings indicate that about sixty percent of the U.S. population are visual learners, who learn best by reading books or through television or other visual media. Another twenty percent learn best when they can move around, as when doing manual labor or being in a workshop. Most of us are a mix. The teaching of Therapeutic Touch covers a wide spectrum of learning—experiential, cognitive, intuitive, transpersonal, and frequently transverbal—and so the teacher of Therapeutic Touch must have a wide variety of skills at hand to match varied learning styles.

The focus of *Therapeutic Touch as Transpersonal Healing* is to bring together the learning process and the healing process. There are several striking commonalities. For instance, both are conscious human processes, and both the teacher and the healer are energizing someone who is in need—the learner in the first case, the healee in the second. There are many ways of learning, and a brief look at them will begin to give an idea of the wide scope of strategies that can be used to teach Therapeutic Touch.

The Use of Presence as a Learning Tool

Possibly the most direct way of conveying the process of Therapeutic Touch is through the presence of the TT therapist. This can be perceived by the student most powerfully when an authoritative person, such as a teacher who is demonstrating TT, and the interested student are both sensitive to the dynamics of the healing interaction as it progresses. As the teacher, in the role of experienced healer, centers deeply and comes into close liaison with her inner self, there is a felt shift in the TT therapist's vital-energy field that may also noticeably alter the local ambience in which she is healing. It is not unusual for a spontaneous hush and individual psychomotor quieting to fall over the observers at that time as well. It often seems that as the healer focuses more intently on the interactive flow of vital energies between herself and the healee, recognizes the disparity of pranic flow between them, and seeks out the subtle sources of the imbalances in the healee's vital-energy field, there is a considerable intelligent but non-physical cognitive force at work that makes itself known not only to the healer and the healee, but also to the observer. In that moment the bystander and the healer can be caught up in an ongoing nonverbal communication about the healing interaction. It is as though an in-depth identification occurs between the vigilant bystander and the practicing healer at that timeless moment. The depth of this modeling is largely dependent, I believe, on the conscious linkage in place between each person and her individual inner self.

It is also in that indescribable place of immediacy and "hereness" that the healer is present for the healee, and so the moment becomes a non-physical agora, or an n-dimensional forum, where subtle information may be exchanged and cognized. The success of this communication depends on the teacher-healer's ability to expand her vital-energy field consciously to include the observer; fortunately, most healers' vital-energy fields are unusually pliant. This ability to stretch the vital-energy field at will is an important factor in exerting presence or charisma, which may be put into play consciously or unconsciously. While the vital-energy field is reactive to appropriate circumstances, it also responds sensitively to intentionality, so the teacher can learn to use this often unrecognized teaching tool upon demand.

In the presence of a teacher who can consciously use this mind function, the observer's perception of the healing interaction as process can be a powerful learning experience. Even if the perception is not fully realized, the observing student seems to get a conceptual grasp of the finespun dynamics involved in the complexity of the healing moment, and this understanding can act itself out in the student as an interior modeling of the TT process. However, the teacher-healer should be aware that this kind of mind-to-mind teaching carries with it a powerful psychic sensitivity that demands a conscious willingness to be open and, therefore, vulnerable. Nevertheless, it often happens that the teacher-healer who is motivated by beneficent and altruistic reasons is most often naturally safeguarded in strange synchronistic ways that baffle intellectual analysis.

An interesting aside to this method of teaching is that not infrequently the healee himself may be caught up in the spirit of this high-level teaching process. Most usually because Therapeutic Touch has helped him, he feels moved to learn how to perform it himself, and may in the future become a student of Therapeutic Touch out of a desire to help others as he has been helped. A system called peer therapeutics is actively practiced to engage the interested healee in the learning of Therapeutic Touch. As he practices newly learned skills on others in need, frequently the learning process itself helps the healee-now-healer to gain insight into his own previous state of illness, often resulting in profound and constructive changes within the former healee.

The Use of Pattern Recognition

The TT assessment combines several cognitive functions: perceptions of subtle energy flow patterns, intellectual evaluation, body language, hunches, and, on a good day, intuitions and clear visualizations. These come together in a kind of pattern recognition that takes place as we pass the hand chakras through the healee's vital-energy field. Crucial to this phenomenon is the ability to perceive wholes out of a collection of "pieces" or bytes of data. In tracking this process in myself, it seems that I get the intuition first in a flash of felt knowing, which is immediately clarified by a clear visualization of that objective knowledge, by symbol or in pictorial form. Pattern recognition helps me relate the various aspects of the experience to each other, so I will better understand what the cues are telling me.

With surprisingly little practice, students can become quite efficient at exerting this mode of perception. Different life forms, such as cats, dogs, horses, birds, etc. are perceived as having distinct vital-energy patterns not quite like those of human beings; therefore, their imbalances may be in different parts of the vital-energy field than one would expect. One way of teaching the student how to discriminate between the various cues is by providing him or her with several opportunities at workshops to access different problems, people, and a variety of vital objects, such as grains, flowers, trees, and small animals that might be available. Be creative. At this writing, one group of steelworkers learning Therapeutic Touch in Pittsburgh practices the TT assessment by trying to determine whether clutches of ostrich eggs are fertile!

A Useful Workshop Format

The most useful workshop format for assuring that each participant assesses several people is to divide the group into smaller groups of four people each, and then divide the smaller groups into two pairs. After centering, the pairs do an assessment on each other, then switch partners with the other couple. When they have assessed their new partners, they change partners a third time and perform a third assessment.

This arrangement works most interestingly for the participants if no information is passed between the pairs until all the assessments are completed. When this point has been reached, the small groups discuss all their findings among themselves, while one member records significant aspects of their experiences. After a break, the entire group reconvenes, and the scribes from each of the small groups report on their groups' experiences. In this manner, every workshop participant has a chance to assess at least three persons—which should provide everyone with a rich experience in the many subtle differences between people as well as a good opportunity to learn about other participants' experiences.

Intuition as a Teaching Tool

Valid sources of intuition can be a valuable source of information about the TT process. Here we must exercise considerable caution because of the highly personal and subjective nature of intuition. However, impressions derived from true intuitions (unlike mere hunches) are implacable in their insistence that the facts are as they are represented to be, even in the face of logic and common sense. In spite of this insistence from one's inner voice, however, it behooves the person teaching about TT from an intuitive base to test the intuition against consensual reality before presenting the material.

The teacher who has honed to a fine edge her ability to access intuition will frequently present her teachings in a way that intrigues the perceptive student, because it may reach beyond commonsense reality, challenging the student to search out a subject more fully and more deeply. Because the nature of the TT process can involve one deeply in non-physical reality, the conscientious teacher accepts the responsibility to help such a student keep him- or herself solidly grounded and stable when operating in situations defined by consensual reality.

Mind-to-Mind: An Ancient Way of Teaching

Still another avenue of teaching Therapeutic Touch is one that is very ancient and very natural, requiring no special equipment or technologies. In fact, many persons—notably those under stress to clearly convey a message to another person at a distance—are able to perform mind-to-mind communication naturally, without formal instruction.

Mind-to-mind communication is a method of intentionally transmitting ideas, information, or clearly defined feelings to another person without speaking to him or her. It is an aspect of conscious telepathy that is based on the ability to visualize clearly and with intentionality the message one wishes to convey. It is a natural means of communication, often used unconsciously. In Therapeutic Touch it is most often used with specific intent during sessions with persons in panic or in traumatic situations. It is also used to reassure and give support to persons who are in final transition, to those who are out of touch with reality, such as psychotics and persons with Alzheimer's disease, and to frightened children. Teachers most often use mind-to-mind communication with their students to convey material that is abstract or beyond the students' experience, or to lend impact to material as it is being presented.

Sensitivity to mind-to-mind communication naturally and effortlessly increases as one progresses in depth in the practice of Therapeutic Touch. The therapist, after all, is frequently working with non-physical energies in the rebalancing of the healee, and is visualizing invisible patterns of vital-energy flow during the TT assessment and reassessment. Therefore, it should come as no surprise that she should finally succeed in her awareness of this telepathic avenue of communication while she is mindfully centered.

As a mode of teaching, mind-to-mind communication should be engaged in either with considerable mature discrimination or not at all. There are more questions than answers about the psychic intricacies of this means of communication, and so one should be respectful of its use. It should not be used at all with new students, for instance. Used appropriately, however, mind-to-mind teaching has come down through the centuries as a teaching tool of distinction, and is particularly relevant in our time of creative revolutions in all means of communication.

Of course, the ways in which each of us reaches into liaison with our inner self is individual, and many people devise personal rituals as aids in this process. When I wish to use mind-to-mind teaching, I begin by centering my consciousness deeply. Then, when I feel in touch with my own inner self (by perceiving the indicators mentioned in Chapter XI) I try to clearly visualize the person—for example, a TT therapist in another location—with whom I want to communicate. Once I am assured that I am visualizing that person with clarity—where he is at that moment, what he is wearing, etc., all details I'll check later to assure the validity of my visualizations—I then communicate with the person through that visual perception. It seems that some aspect of my eyesight (possibly the ajna chakra?) is the structural medium for the communication, for there is a different dimensionality to the depth of what I am "seeing." While this process is going on, I seem to be able to perceive objects in the healee's vicinity through the eyes of the person with whom I am communicating, and who is actually observing or thinking about the healee. Later, I check with the healee to determine to what degree my perception matched his impression of the experience.

The teacher of Therapeutic Touch will choose which of these styles are appropriate to her students' abilities and to the material she is presenting. An important decision will be based on the background, experience, and mindset of the students. Because there are admittedly still large areas of ignorance and misconception about the vital-energy field and the part it plays in the health or illness of living beings, much of the teaching of Therapeutic Touch could be called exploratory, involving the use of analogies, the playing of games, the development of learning models, the exercise of Therapeutic Touch itself, and the use of theory building and creativity in forging new learning tools for contemporary students whose understanding of subtle energies is hovering on the cusp of a new time and a new perception.

Expectations of TT Sessions

In order to have realistic expectations for Therapeutic Touch, it is helpful for students to understand certain concepts. First, they should be aware of the many connotations of the word "heal":

alleviate
diagnose
harmonize
improve quality of life
make healthy
make whole
mend
rebuild
recuperate
regenerate
remedy
renew
restore
revivify

Students should also recognize that healing sometimes works in contradiction to accepted knowledgeable predictions, and that the reason healing may work for one person and not for another under similar conditions remains little understood. Qualitative factors, such as religious and social belief systems, personal values, personal goals, and personal meanings given to illness makes the simple summing up of multiple factors untidy, yet they can be critical properties of the healing process. The TT therapist may interface her sessions with other therapeutic services to insure that crucial concerns of the healee's healing process are subsumed within a holistic context.

Besides psychosocial and neurophysiological referrals, several other optimal therapies may be appropriate; those that have been found to interface most readily with Therapeutic Touch and to afford the healee the most collateral benefits are acupuncture, aromatherapy, imagery (as "homework" for the healee), and Swedish massage. There are several self-help strategies that also are useful in conjunction with TT sessions, such as meditation, tai ch'i, individually designed hydrotherapies or aerobic exercises, sauna, jacuzzi, and hot mineral springs.

Students should be familiar with the conditions that respond best to Therapeutic Touch. The system that is most sensitive to Therapeutic Touch is the autonomic nervous system (ANS). Because of this, Therapeutic Touch is most useful with psychosomatic illness, which accounts for approximately seventy percent of health problems in the world. Hardly less sensitive are the lymphatic and circulatory systems. Problems of the genitourinary and musculoskeletal systems, particularly bone fractures, also respond very well to Therapeutic Touch.

There are several systems in which treatment by Therapeutic Touch is only partially successful, such as the collagen system: while arthritis responds well to Therapeutic Touch, there are mixed findings regarding lupus. The endocrine system also has mixed results: Therapeutic Touch is very effective for problems of the adrenals or the thyroid, but is not remarkable for pituitary dysfunctions. Therapeutic Touch works well with certain problems of the reproductive system, such as problems of pregnancy; however, it does not do well with dysfunctions of the pancreas, such as diabetes. While Therapeutic Touch has never healed persons with completely severed spinal cords, several such patients (with fully transected spinal cord) report that they can feel specific areas where the TT therapist is working in their vital-energy field without the therapist making skin contact or exerting pressure of any kind. Therapeutic Touch also has had significant results in cases of manic-depressive illness and in people who are catatonic or in a coma; however, other psychological disorders, such as schizophrenia, have not been successfully treated with Therapeutic Touch.

The several types of illnesses that will evoke the fullest benefit from Therapeutic Touch include:

1. Those due to mechanical problems, which leave scars or dysfunctions, as in the case of bone fractures, wound healing, or blister or pustular formation
2. Infections, in which one seeks to change the patterning of the immunological response
3. Stress-related illnesses, whose fundamental problems are concerned with behavioral and attitudinal change
4. Illness due to side effects of prescription drugs or treatment. In 1999, iatrogenic medicine (*iantros*: physician, *genic*: produced by) was the second largest cause of death among hospitalized patients in the United States. Because of the heavy reliance on high-tech pharmaceuticals whose subtle ramifications continue to be little understood, an inordinate amount of allopathic medicine is actually iatrogenic medicine. Infamous among these are steroid medications. Iatrogenic effects for which Therapeutic Touch has been most helpful are nausea, vomiting, radiation burns, and problems stemming from fluid and electrolyte imbalance. However, retaining the state of balance achieved with Therapeutic Touch may be difficult, particularly if the patient continues to take the medication that is causing the problems.
5. Illness of spirit, a condition little recognized in the West except in religious therapeutic practices, but acknowledged in all cultures that advocate active shamanic practices. Illness

of spirit is demonstrated by lack of meaning in one's life, loss of purpose, loss of self-confidence, or loss of personal values.

6. Finally, TT works very well in rebalancing all four traditional vital signs—pulse, respiration, blood pressure, and temperature—as well as the most recent symptom to be proposed as a vital sign, pain.

Like most healing modalities, Therapeutic Touch works best with acute illness of recent derivation, and works least well with genetically derived illnesses and in cases of denial of illness or hostility toward the TT therapist. And, like all healing modalities, the TT process is bolstered by social acceptance of the healee and his adaptation to or compensation for the dysfunction caused by his illness.

Based on clinical studies and basic research, the anticipated results of Therapeutic Touch interventions with the highest reliability are:

1. Rapid relaxation response, usually occurring within two to four minutes and consisting of reduced blood pressure, respirations, and pulse, and an increased dilation of vessels, particularly in the peripheral vascular system
2. Reduction of pain, demonstrated objectively by a concomitant decrease in analgesics and an enhancement of their effects
3. Acceleration of physical and psychological healing
4. Reduced anxiety, e.g., relief from nausea and enhanced respiration

Therapeutic Touch also promotes a positive change in emotional affect, increased effectiveness of systems management (e.g., the absorption of medication, the therapeutic effects of chemotherapy, etc., in patients with AIDS and cancer), and a peaceful final transition.

Of course, the body has a fantastic ability to recover. We call it "maintaining the integrity of the system," or recuperation, which is a process of self-healing by any definition, so that in the final analysis it is the healees themselves who are the significant arbiters of their own healing. We as therapists help, support, and quicken that life-affirmative impetus, but it is the integral being of the healee that accepts and actually permits that healing to occur.

Clarifying the TT Process through Discussions and Insight Questions

In the following pages are several discussion topics and insight questions for use in Therapeutic Touch seminars or other workshop courses. It is suggested that the reader also replicate or use as a model for Therapeutic Touch exercises that have been previously published.[1] The topics and exercises included in this book can be used as they are, if you are teaching or practicing Therapeutic Touch, or you can use them as a model to develop your own. It has always been the practice in our teachings on Therapeutic Touch to encourage students to develop their own creativity. However, please keep in mind that unless you maintain the principles Dora Kunz and I have enunciated as we developed and clarified the TT process, you will not be teaching or practicing Therapeutic Touch.

The most interesting way to start a discussion on the healing process is to ask a question, or to challenge accepted assumptions. Then—most importantly—allow the students to respond as fully as they are able. Let everybody who wishes to speak do so.

Encourage dialogue between the students, but ask for clear, pragmatic examples to support the discussant's points. The fact that Therapeutic Touch is a transpersonal process makes it imperative that the therapist be well grounded, while having the courage to actualize the many realities to which her consciousness holds the key.

Make no comment until all have had a chance to air their salient views, but take notes on the students' comments, using verbatim quotes when possible. From time to time glance at your notes and try to find opportunities to cross-relate the students' remarks. If they agree, encourage them to figure out whether they are arriving at their conclusions from the same assumptions. If they disagree with each other, challenge them to see if taking one step further might lead to a reconciliation of ideas, or to the discovery of a new line of thought, or to a reconsideration of their positions.

You yourself needn't have the answer; rather, your role as teacher is to focus attention, to help articulate the important questions, and to help students leave the discussion room enthusiastic about the opportunity for a creative exchange of ideas. Below are several suggestions for discussion topics about the healing process and about TT.

The whole person

How do you visualize treating the "whole person"? Is there a way you can be assured you have reached that level of consciousness in the healee?

Levels of energy

Illness may occur on many different levels of subtle energies. Write one paragraph, as the basis for later discussion, on three ways you know what level of energy you are dealing with as you do a TT assessment.

Human energy

You have to be willing to study your own energies so that you can recognize how to use them to benefit someone else. To understand the power healing can have in your life, answer the following questions, writing the answers in your journal. Where you are asked to indicate the level of intensity of your feeling, use a scale of one to five, with five being the highest level of intensity.

- At this moment, how would you rate your energy level?
- Can you sense where its focus is?
- If there is an emotion you are feeling at this time, how would you describe its energetic characteristics?
- What thoughts are passing through your mind at this moment? What effects do you think they will have on yourself and others? How would you make them more effectual?

Love

Brand reports that studies have demonstrated that love is relevant to successful healing and to healing analog studies (experiments that measure the effects of distance healing).[2] Findings indicate that the healer's feeling of love for the healee plus strong, positive feelings of merging and interconnectedness may facilitate successful outcomes. Do you use love as a therapeutic medium during your TT sessions? What does it feel like when you send thoughts of love? How do you know it is love you are sending (and not a less intense feeling, such as affection)? Do the "vibes" differ depending upon the person to whom you are directing the feeling of love? Can you send love to someone you really don't like? How do you use love therapeutically?

Compassion

Think of a situation that would evoke a sense of compassion in you, such as a tragic occurrence reported in a recent television newscast. As you recall the incident, try to visualize clearly one of the persons involved in the tragedy, perhaps the victim or his or her relative.

Do you feel a decided shift in your energies as you think of this person? Record in your journal the intensity of this felt shift on a one-to-five scale of increasing intensity, and describe the internal movement of that shift.

Did that shift have an emotion related to it? How would you describe that experience?

Nonlocality

Have you sent thoughts of help or healing at a distance to the person described above? If you have, describe what you did in as much detail as you remember. If you haven't and wish to send such thoughts, do it now. Record your experience in your journal.

Compare your experience with the thoughts on compassion in Chapter VIII. What new information about yourself have you learned? Record your insights in your journal as a basis for group discussion.

Pain

What does pain feel like? Do you feel it in a locality? a region around you? in your body? Are there perceived degrees of intensity? Are your chakras affected a) as the healee? b) as the healer? Can you "turn off" feelings of pain? How do you do that? If not, how do you endure pain? Can you describe these responses?

Intuition

Intuition has been called a dialogue of one's inner self with the inner self of others. Where do you "hear" your intuitions? How do you suggest one go about increasing one's sensitivity? How do you learn to separate out fantasy from intuition? Is it possible to add depth to intuition, e.g., to get more information or more explicit data from the experience, or to go more deeply into your insight? How do you separate out emotions from intuition or true insight during the TT assessment?

What signs tell you that you are trying to get an intuitive grasp of a situation? How do you maintain clarity when attempting to be intuitive on demand? Are there subjects about which you are more sensitive than other subjects? How much credence do you give to your intuition? Do you have personal criteria for the acceptance of intuitions that arise in yourself? in others?

Exercises

Empathy and Intuition

- Choose a position in which your body is comfortable, and center your consciousness.
- Breathe slowly and fully for a few moments, noticing the relationship between the rhythm of your breathing and pulsations or feelings of sensitivity to your own vital-energy field.
- Visualize the person, animal, bird, plant, tree, or object of interest; then extend that visualization so that the person or object appears to be within your vital-energy field.
- Closely examine the visualization and then try to shift your attention so that you identify closely with that object and you feel you can understand things from its point of view.
- What is that object saying to you? How do you respond? Are there any surprises in what this other perspective conveys to you? To what extent is this interchange "real"? Did you obtain any new information during this exercise? How will you use it?

The Inner Self

As one progresses in depth in the integrated use of Therapeutic Touch in one's lifeway, that person can become more cognizant of the importance of recognizing one's rootedness in a spiritual nature. Has this happened to you? Are you aware of a deeply felt presence or higher state of consciousness that is there for you when you permit it access to your attention? If so, talk to it for a moment. If not, quietly "listen." How would you describe the nature of this inner self? If you feel that connection might be useful to you, how do you think one could maintain a closer alliance with the inner self?

Having experienced the Therapeutic Touch Process interaction, what do you think are the most important points to include in the teaching of future students of TT about the following phenomena?

a) The subjective connecting and reflecting that occurs within the TT therapist herself
b) The cues, flows and rhythms that arise in the healee's vital-energy field and become cognizant to the TT therapist
c) Implication of these cues
d) How the cues are rebalanced

Physiological Effects of the TT Process

When a TT session is observed, the ambience is usually quiet and relaxed, the movements are gentle as the TT therapist moves her hand chakras through the healee's field, and very little seems to be going on. Nevertheless, if the treatment is successful, there are several body processes that are significantly affected as the TT therapist tries to help the healee reestablish harmony, including:

cellular function
blood circulation
fluid and electrolyte balance
hormonal tides
metabolism
nerve impulses
respiration
waste product excretion

How would you account for so much happening with so little physical effort? What do you see as the significant inferences one can draw from this occurrence? Are there precautions one therefore should keep in mind?

Placebo Effect

In small discussion groups of four to five persons each, consider the various things you do during a TT session that contributes toward a placebo effect in the healee (i.e., anticipation, suggestion, yourself as an authority figure, etc.). Report the group's comments later to the rest of the class.

Personal Growth

Ask yourself: What is my role in the universe? How am I fulfilling that role? What archetypes speak to me? What are the bases of my own assumptions about my role? What do I think I might learn from my role in the next three years?

- How did you get involved with Therapeutic Touch? What made it happen for you? What is there about Therapeutic Touch that made you realize that you could use it as a significant part of your lifestyle?
- Do you know why you want to heal? Now is a good time to confront this question, either in discussion with friends or by writing a paragraph or two in your journal. Take a few moments now and do it. Ask yourself: Why do I want to be a healer? There are no absolutely right or wrong answers, but you should understand your own motivation.
- As a test object for the universal healing field, what characteristics of that field does the TT therapist convey? Consider the following answers and discuss:
    - A sense of stillness and peace
    - Feeling at ease with other than the three dimensional world
    - Compassion

- Knowledgeable use of the chakras as a medium of communication
- Sensitivity
- Being unattached to the results of the TT process
- Willingness to be a human support system
- Willingness to explore the deep self

Therapeutic Touch and Personal Growth

The Message From Yriaf: A Fantastic Guided Imagery

Equipment: Several slips of paper and a small bag. Copy the messages below onto the slips of paper and put the slips in the bag. Use as many as you wish, or make up your own imageries of transpersonal experience.

- You regard your TT practice as a personal growth experience.
- The cues you pick up from the vital-energy field during Therapeutic Touch are very important and meaningful to you.
- In the stillness of your meditation, you know the answers.
- Often it is only after you have finished a TT session that you realize that the experience had a timeless quality to it.
- There have been times when you traveled to a new region of the world that you've had lucid dreams of ancestral experiences.
- Often when you see someone in dire circumstances, compassion wells up and the context of the situation changes.
- Centering acts as an empowering agent in your life.
- While you are in a centered state of consciousness, you can become aware of the functioning of several of your chakras.
- As you assess the healee's vital-energy field, you "listen" for shifts in consciousness to inform you of imbalances in the field.

- Responses to your aspirations toward your inner self are very real to you.
- Sometimes, during routine activities, you are aware of familiar racial memories going through your mind.
- In the depth and constancy of your meditations, multiple realities become familiar to you, as though you always had known of them.
- Upon awakening from a deep sleep, for a moment you have had a clear sense of having had contact with other kinds of consciousness.

Instructions:

- Settle down comfortably and close your eyes. Take one or two deep breaths and give your body permission to relax. I'll give you one or two minutes to do this.
- Now, without opening your eyes, glance over your shoulder and notice a diminutive fairy shimmering in the bright light of the sun. The fairy apparently has entered the room through a crack in the window.
- Now the little personage is flying around the room. Watch it as it alights on your arm and smiles. Captivated, you smile back.

  The fairy leans toward you and says, "Hello. My name is Yriaf. I come from a far land, and I have a message for you."

  "Me?" You say in surprise.

  "Yes, you. It comes from my friend, who also would be your friend. Think about the meaning of the message. I shall return."

  And so saying, the little fairy, Yriaf, motions to the assistant, who will bring you a bag.

- Open your eyes and wait quietly until the assistant gets to you. When the assistant is before you, reach into the bag and select one of the paper slips.
- Read your message. Don't tell anyone what it says. Put the message in a safe place. Refer to it several times, including just before you go to bed.
- Journal the effect the message has on you, noting any effect it has had on your relationships, circumstances, or how you view the world during the next several hours/days. You will have an opportunity to discuss your experience in our next seminar/discussion/workshop.

(Note: The name Yriaf [I-ree-af] spells the word "fairy" backwards.)

Insight Questions: The Care and Feeding of the Grounded Intuition

Healing and Creativity

The healing act and the creative act are very closely allied. To be creative, one must have permission to use the imagination.

- How do you assure that you do not let fantasy, personal projection, or bias unduly influence your TT assessment and interventions?
- What are your personal touchstones of reality during TT sessions?

Encouraging Self-Healing

Therapeutic Touch empowers the healee to experience his or her own natural self-healing abilities in conjunction with the support given by the TT therapist during the healing interaction, and increasingly over time as the healee's immunological system becomes more effective.

- How do you ensure that the healee makes the most of this opportunity?
- What do you tell the healee?
- What homework do you give him?
- How do you check up?
- Cite one case history.

The Therapeutic Touch Differential Assessment

The effects of the TT process can be dramatic or subtle. What are the differences in cues that you pick up, and how do you intervene differently, with healees with the following problems?

Physical

a) Rheumatoid arthritis and low back pain
   - From your experiences with Therapeutic Touch, what do you think would be the significant differences and similarities in vital-energy field cues that you would sense during the TT assessment and rebalancing phases with healees who have rheumatoid arthritis as compared with those who have low back pain?
   - Can you cite a case history that exemplifies these differences?

b) Severe anxiety and hypertension
   - From your experiences with Therapeutic Touch, what do you think would be the significant differences and similarities in vital-energy field cues that you would sense during the TT assessment and rebalancing phases with healees who have severe anxiety as compared with those who have hypertension?
   - Can you cite a case history that exemplifies these differences?

c) Deep depression and bereavement

- From your experience with Therapeutic Touch, what do you think would be the significant differences and similarities in vital-energy field cues that you would sense during the TT assessment and rebalancing phases with healees who are severely depressed as compared with those who are experiencing bereavement?
- Can you cite a case history that exemplifies these differences?

d) Cancer and AIDS
- From your experience with Therapeutic Touch, what do you think would be the significant differences and similarities in vital-energy field cues that you would sense during the TT assessment and rebalancing phases with healees who have cancer as compared with those who are HIV+?
- Can you cite a case history that exemplifies these differences?

Emotional

a) Restlessness and agitation
- From your experience with Therapeutic Touch, what do you think would be the significant differences and similarities in vital-energy field cues that you would sense during the TT assessment and rebalancing phases with healees who are restless as compared with those who are agitated?
- Can you cite a case history that exemplifies these differences?

b) Rage and anger
- From your experience with Therapeutic Touch, what do you think would be the significant differences and similarities in vital-energy field cues that you would sense

during the TT assessment and rebalancing phases with healees who are angry as compared with those who are filled with rage?

- Can you cite a case history that exemplifies these differences?

c) Concern and anxiety

- From your experience with Therapeutic Touch, what do you think would be the significant differences and similarities in vital-energy field cues that you would sense during the TT assessment and rebalancing phases with healees who are concerned as compared with those who are anxious?
- Can you cite a case history that exemplifies these differences?

d) Indifference and lethargy

- From your experience with Therapeutic Touch, what do you think would be the significant differences and similarities in vital-energy field cues that you would sense during the TT assessment and rebalancing phases with healees who are indifferent as compared with those who are lethargic?
- Can you cite a case history that exemplifies these differences?

Note: Using this same process, consider mental aberrations such as:

- Dissociation and contemplation
- Psychicism and spirituality
- Hunch and intuition
- Creativity and fantasy

Working with Other Methods of Healing

TT can interface with many other healing modalities—contemporary, traditional, and alternative—such as:

massage	progressive relaxation	chiropractic
biofeedback	meditation	Trager
deep shiatsu	guided imagery	naturopathy
clinical hypnosis	herbal essences/teas	prayer
allopathic medicine	professional nursing	

In general, what applicable principles underlying which one modality would you think might most benefit the TT therapist? How can we learn from them?

Real-Time Transcript of Workshop on Advanced Therapeutic Touch

Note: Below is a minimally edited transcript of a workshop I gave for advanced TT therapists. I have chosen the verbatim text because I think it will give the reader a true picture of the thinking of the student as the TT workshop proceeds as well as an idea of the various techniques the teacher needs at her fingertips to integrate a diversity of views and experiences so that the ongoing discussion provides a consistent, reasoned, and coherent medium for the continued learning of all the participants of the seminar.

In the body of the transcript you find reference to a series of photos of TT sessions. To follow that discussion, please refer to the photos on pp. 128–138.

Dee (D): What I would like to do tonight is somewhat experiential. I was thinking about this while watching you at this morning's clinical session. As I looked around, I realized that you were doing Therapeutic Touch (TT) in four different ways. I quickly sketched these different pos-

tures, which you see below. Do they look familiar? (laughter) The thing I want to ask is, do you see yourself in any of these sketches, and would you mind talking about what you were doing when Krieger's Camera Eye caught you?

1. One Healer doing Therapeutic Touch to Healee

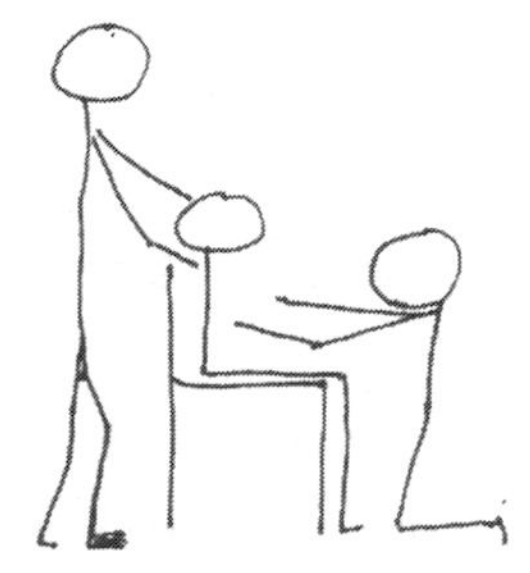

2. Two Healers doing Therapeutic Touch to Healee

3. Two Healers and Healee joining hands and centering before doing Therapeutic Touch

4. Healer doing Therapeutic Touch to Healee, who is on a massage table

Audience (A): Yes, I recognize my slim waist!
D: Who was working on the massage table? Maria?
A: I work on the table a good deal because I am a Trager practitioner and I do TT on the patient before every Trager session. My clients are already on the table, prepared for Trager.
D: Okay, and why do you do the TT before it?

A: It gives me more information about what's going on; it lets me put the physical energetics together. It also centers me before I do the Trager work, although Trager has its own way of getting in touch. It's called hook-up.

D: So tell me more about centering. What does it do for you?

A: Centering brings me more fully present in the moment.

D: Yes, and . . .

A: It allows me to mobilize more of my own inner self and more of the inner skills that I have; it brings that all to one point, into focus. It opens me up.

D: It sensitizes you?

A: Yes, it opens me as a channel for the healing energy, and in Trager work there is an element of that; that's why the two are so wonderful together. There's an element of energy moving through me while I'm doing the Trager work, so centering opens the channels for that.

D: Hmm. . . .

A: The thing it really does for me—and you may call this TT or not—it makes the "work" of the Trager work more sacred.

D: Okay, now you've all bought into it. (Laughter.) I can do this to Maria, she's been a great friend and will understand when she sees where I'm going. So, the question is: How do you know?

A: How do I know which part?

D: Yes, the latter part, the whole experiential thing that you were talking about. The energy passing through you, that it made you feel more sacred. What were your "touch" experiences, how did you know what was happening?

A: Right.

D: How you do know, for instance, that you just hadn't gone to sleep and you were having a hypnotic vision, or something?

A: Yes. My consciousness of the task at hand shifts.

D: Okay, good.

A: Instead of me putting my hands on a patient and doing Therapeutic Touch and then doing Trager, I become aware that I am doing something deeper than that, and I know I'm aware of that because my awareness of all the different levels of myself and my client becomes . . . suddenly they are in front of me. I've shifted from a day-to-day view of the healee's body as a body, and mine as mine, to a place of seeing how they both fit together. Suddenly I am aware of all those potential ways that living things can fit together and I begin to see connections between what is happening physically, emotionally and spiritually, mentally, whatever. So that I am interacting with the client in a way that is very different than the way I do in hospice work, for instance.

D: What do you mean by "different"?

A: In hospice I walk into the family's house and I'm very clinical, medical-modeled in most cases. I take the vital signs, note the symptoms, how the patient is coping; I'm looking at all that "stuff." But when I shift to a centered place, I'm suddenly acutely aware of how . . . actually I pick up more and because I'm in the moment and I "see" subtle ways that the body language expresses what is happening about things I actually can't see, emotional feelings and that sort of thing. I become much more intuitive, although that also leads in my hospice work now. I become much more intuitive in my Trager work than I would be in my normal workday world. I get information in ways that I don't always understand, that I'm not usually accessible to, like when I'm driving my car, or whatever, doing day-to-day activities.

D: Does that make sense to anybody?

A: Yes.

D: Okay, why? What is there in you that makes sense? Did you have similar experiences?

A: Hmm . . .

D: Somebody speak up. (Laughter.) Yes, go for it, Cathy. By the way which of these models are you, number 1, 2, 3, or 4?
A: I'd say I'm more with the person on the chair, I think.
D: Okay.
A: I'm the person behind the healee.
D: Okay.
A: I can resonate with what Maria was saying because much of what she said was similar to what I've experienced when I'm working with someone.
D: Is there anything particular that she said that resonated very highly with you in a way you could articulate?
A: Yes . . . when she spoke of the centeredness and being more in an awareness of the present moment. But it's also a deeper awareness of myself, my inner self . . . I sometimes call it my true self, in a way, by coming to a still point within me, and then resonating or relating more to that truer or inner self in that person whom I am working with.
D: Okay, and how do you know that for yourself? What are your touchstones that tell you this experience is valid, that it is a "reality"?
A: My touchstones are a deepening within myself, my own being. I feel that the other things in time and place . . . the fields, they are "someplace," but there's an inner consciousness from within myself that I'm much more present to, and in that moment, I'm much more there. It's that inner presence, inner self that's relating to that person that I'm touching or assessing. It's much different, it's much deeper and the connections are different—as Maria was saying—than in my everyday life, although from doing TT so long, that's happening more and more.
D: Okay. Yes?
A: I'm 3.
D: Okay.
A: I agree with everything Maria and Cathy said, and also something that occurred to me that happens when I do that. [. . . Y]ou go into

this very still, deep place within you, but there's something about that process that becomes expanded and connects with a layer or level of experience that isn't accessible when you are not in that state. It's not as successful.

D: How do you prove to yourself that you are not just hallucinating?

A: Because of the level of attention and presence.

D: Good. Can you explain that a little more?

A: This is just my experience, but I would think that if I were hallucinating, there would be more of a story I would be telling myself about what I was experiencing. And so, to simply be present and to be still and observant and sensing I think is different than creating a story about what you think you might be feeling: you are absolutely in that moment.

D: Did you want to say something?

A: Yes. I am 3. From my own experience, the reason that the kinds of experiences that have been stated—about the timelessness and the being still within and the knowing that something not physical is real—are, I believe, so, and I'm probably not hallucinating, is that part of my experience has been, when it's appropriate to have that, to know I've had that experience before. Some of the time what I have experienced and what the other person experienced matched; they were the same or similar and then I know that communication took place. Experiences like that, over time, prove to me that it was real.

D: Jody, do you have something to add? Do you agree with that?

A: Often when I'm at work doing TT and I think I feel energy more, or that it's a pain spot, then I will ask the healee: Is it sore here? So I can test what I think I am feeling with my client. And so I get that practical feedback, too.

D: Okay. You also wanted to say something?

A: I'm trying to put words to my experience. Especially in the last year, when my TT treatment is going on, there is a point where I seem to step into a flow, like a stream of energy, and I as me, myself, dissolve. I have control of my intentionality, but there is not so much a sense of me as—I don't know how to put this—it's like I'm just participating with this flow of energy and my participation is the focus of the intentionality, but there is not so much of a sense of self in it.

D: How do you know you are not prevaricating?

A: Because I feel like I'm really participating in what is going on.

D: You are identified with it, is that what you are trying to say?

A: Identify . . . it's that I'm fully aware of where the energy is going and what is happening with it and where there are interruptions in the energy flow. So I feel like I'm fully participating in this energy movement, this happening. It's that that makes me feel I'm not dissociating.

D: Okay, all right. Anybody else? Yes, go for it.

A: For me it's a matter of being detached from the experience, rather than being dissociated. Detached is a . . . I had it a minute ago. . . . (Laughter.) Detached is being very present and watching. It's the words, they are so difficult.

D: Say it in Australian. (Laughter.)

A: Detached is a . . . . Dissociated is being disengaged and watching, whereas detached is being engaged and watching.

D: Good! You get an automatic "A" for the course! That was very well done. How about anybody else? Go for it.

A: I do TT like model 3.

D: This one? Does anybody else recognize themselves here?

A: I think I do, I'm not sure.

D: There were three people and they all were holding hands as they formed a triangle and, I suspect, they were centering.

A: Yes.

D: Was that you?

A: Yes. That was a really unique experience for me. I haven't done that before, in that way.

D: Who was with you in the group, by the way?

A: Toby and I were doing a treatment on Nancy, so the three of us centered together initially. Then we went to the second model, the more traditional, you might say. But it was unique in that I felt we had full cooperation and understanding between us. It was like we had a deep connection between the three of us that was somehow different.

D: Do I remember correctly? Was it Nancy who asked to be part of it?

A: She had told us that she did some energy work and wanted to be part of it. It seemed to work very well, I think. However, it wasn't until I saw the sketch that I realized that we had put it into a form.

D: All right. Toby, what was your experience of that?

A: I think it gave the healee a sense of full participation in what we were doing. I think it was at that moment that it brought her into equality with us.

D: What was the experience saying to you while you were involved in it? Were you waiting for it to end, or was it bothersome?

A: No, no. I think it was full participation. I don't recall thinking anything other than that at a certain moment we all made eye contact when we knew we were ready to move on.

D: So you had eye contact at the same time?

A: Yes, yes.

D: Something had gone on, something you all participated in, something had happened so that you could all agree: yes, this is the moment and now we go on to the next thing.

A: Yes.

D: A kind of tacit consent among you.

A: Yes.

D: That's very interesting, really great that you had that moment of mutual experience. Yes, you had something to add?

A: I'm identifying with the 3 model, in a sense, and that is in reference to the volunteer TT clinic I run. We start with a group meditation, and recently invited the clients to come early to join us. Then we just move quietly onto the tables after the meditation. We are finding that their responses to the treatment feel much more magnified somehow. There is a sense that they are opening more readily to the treatment and this seems to occur in a very powerful way for them.

D: During the time that you have the meditation, what is your experience?

A: A tremendous sense of connection. A very deep quieting.

D: A quieting, a stillness. That's good.

A: Yes.

D: Anyone else? Yes?

A: As it happened, today the client wanted us to hold hands and one of the things she expressed afterwards was that there was a transition that she experienced the day before with the healers. She said she wasn't aware of the treatment phase, that there didn't seem to be a break in her state of consciousness.

D: Okay. Is there anyone who would like to speak?

A: One of the things that struck me as Maria and Cathy were talking, the word that comes to my mind when I'm working with a client is: authentic. That my authentic self, without any cover story, and the client's authentic self connect with each other.

D: I like your term "cover story." What does that feel like for you?

A: There is the peacefulness that people have identified. A sense of relaxation into the connection and the sense of oneness. And that we, the client and I, are in this together. That there is this authentic, true communication about what is needed and how to respond to that.

D. Good, we're getting there. Yes.

A: These words that you said, true communication, rang a bell with me, and so did the word resonance. I identify with model 1. To me, the quietness allows me a deeper awareness of the whole present moment. In ordinary consciousness I think we may be partially aware of the present moment. In TT there is a deeper awareness of the whole; the moment, the present, is complete. For me the level of trust and genuineness goes along with the sense of being true that develops when I'm doing TT. There's this relationship of deep trust between healer and healee. That level of energy allows for much more abundance of trust.

D: Yes.

A: I find that over the years I do not have many words or pictures, but the TT process is very physical. When I approach the time of healing it's like a Niagara Falls of energy flows through me. It's a very physical thing and I don't do as much strategizing. I just allow that to happen and then I'm more aware of how the energy is moving through the healee's body, but I don't so much maneuver any more, I just more or less know when to quit.

D: Okay; however, how do you know? How do you know the Niagara Falls isn't just an allergy?

A: Good question. Because it seems a direct response to my approaching the healing act. I seem to invite it. And yet it also seems to be conscious and friendly.

D: Okay.

A: Replicability, that will tell you.

D: There you go. Anybody else?

A: With respect to what Molly was saying, first of all, I identify with the person using the table. I've just gotten a table and started using it. . . . I find it easier on my back and there's plenty of room to work. I find, like Molly, that there's less planning and strategy. I'm doing the assessment and get information, but it's not mental.

D: Well, what is it? How do you get that information?

A: I don't know. It's not guessing, but there are no words that go with it.

D: You just know that it's right.

A: Yes, I just know that it's right and I feel that I'm in a slipstream of rightness for me and the healee; the feedback is relative to what it is I've been getting in this odd way. So I feel really trusting. I know, first of all, that I will do no harm. I know that absolutely, I just know it and so I trust that whatever happens is right for whatever that person needs, that at some level it's none of my business.

D: Wow! That is a deep, deep level of trust. Well, who takes responsibility?

A: Well, I feel really trusting that the wisdom that's inside the person that I'm working with is in charge of what is happening. Obviously I am responsible for my behavior, for the setting, etc. I haven't tried to put this into words, but the baseline that all my work comes from is trust.

D: All right. Anybody else tried it that way? Can you help her out?

A: Does that make sense?

D: We are going to find out in a minute; who wants to talk to it?

A: Yes, with my students when I teach I like to get their minds unengaged instead of listening to the monkey chatter in their heads. I tell them that their responsibility is to do the best TT that they are capable of doing, and the rest of it is up to the client.

D: So you put the responsibility on the client.

A: But the TT therapist's response is to be open to all the information that's out there, not just what they are used to seeing . . . and they are also responsible for being vulnerable, that is, open, so that they can receive the information they need to receive from the field. It's very hard to open to that vulnerability that the TT act implies. So, they are responsible for putting themselves out there for that. But beyond that, they have to trust the healing wisdom of the client.

D: Anybody want to respond to that?

A: The more I experience that peacefulness, the more I trust it. There's not harm in it. That trust is what I'm working with in the person. It helps me to react more from my center.

D: I'm not trying to be a wise guy, but do you think it's possible for someone to be peacefully ignorant?

A: Yes. I feel peacefully ignorant a good deal of the time. I mean, it's a continual learning process. I've learned a lot, but there's a heck of a lot I don't know.

D: Anyone else? Julie?

A: It may be just a problem of semantics, but my personal point of view is that to leave decisions up to the wisdom of the client; well, I wonder. This morning we had a client who wanted the TT to go on, although we had been working on her for more than 45 minutes. She had no idea of the possibility of overdose. It just "felt good." I think there has to be a partnership between healer and healee and you have to help the healee understand the implications of what TT is really about. TT is about rebalancing, and there has to be a balance in our relationship as well. Yes, that person is ultimately the healer, but you may have to help them understand the subtle dynamics of what they are into. More is not necessarily better, it could be dangerous if one overdoses even on energy.

D: Okay.

A: What I heard was in reference to their higher self, if you will, or inner self. That aspect of themselves that knows its own wholeness.

D: You want to respond, don't you?

A: I was going to say that in point of fact, in the midst of the flow of being your inner self, and working with the inner self of the client, that's when the information comes that you should shut this down. It's the inner self that would have been telling us to shut down after twenty or thirty minutes instead of forty-five. It's that personality or ego self that is on the surface—which I can identify with, because that is what I have to get past to get into the flow myself. That's why I trust it so much because what is underneath that. . . . I don't want to force spirituality down anybody's throat but I come from a very strong place as a Quaker. Our form of worship is the Silence. And we literally center, and we go in.

When we come to that deep place inside ourselves, we suddenly find out that it is the same deep place inside everyone else in the room, and we are all flowing in the same river. And in that river is the power of whatever you call Spirit. And that is what happens in TT, except we use that power for healing. And that is why the act becomes sacred. We have come to that point of oneness that is the total knowingness—the whole vital energy that runs the universe—inside ourselves, and we have used that to communicate with the client. And if we are both able to make that place, the truth of what is needed is so pure that it is the healing wisdom and we can trust it infallibly.

A: Both of you have responded in ways that really reflect what I feel. I am in charge of the treatment in terms of the time and how it goes, and it is not the personality of the person. The deep trust is at that really, really deep place, and I am the person who is doing the treatment with feedback from the client. But the underlying . . . the deepest place in me knows that the deepest place in them knows what they need. So I can go in with my mind and say, "I really want that broken ankle to feel better, and I want that person's shoulder not to hurt them any more." And what has happened to me a lot is I have had to detach from the outcome: I go with my personality wanting to fix the person's back and the grief from the death of their mother five years ago is what comes, because the wisdom inside them knew what needed healing. It wasn't for me to decide. I could focus my attention on the presenting problem, but with whatever the interaction was . . . however deep it went . . . something that really needed healing would come out. So I've gradually gotten it that on some level it is none of my business what the deepest healing place in them decides to do with this freely offered and freely given energy of healing.

A: What actually arises . . . emotional states . . . unexpected things of different kinds . . . whether it's transient pain in new places . . . or whatever it might be. I feel really humbled when I am entering that healing place with the person. Although I trust that sacred space as such, I hope I have the skills and the wisdom to be there in the right way for that person. I do feel quite a responsibility for that. That belongs to me.

A: I guess my problem with this is that . . . we all know when we do treatments on people that things have arisen. Things that we didn't necessarily pick up on the assessment. But I really feel when I teach that the goal . . . the aim . . . is for people in the assessment to get all that information up front. And so, too, to create the opportunity for that to happen. To center as deeply as we can by connecting with our inner self . . . in that way for each of us.

I think that we really need to be clear about how we do it. We can't replicate something if we don't know what we did. It's hard to get the language. It's hard to figure out what our experience is internally, what it feels like, what kinds of ways we know something is real as opposed to fuzzy. All of us have times in our lives when we just know something. It is very clear, there is no question in our minds, and it is borne out. When we have that experience, there is a particular set of cues that tell us that it is that experience. And to use that as a gauge when we get information in TT assessment and so on in terms of what kind of information is it like. Is it fuzzy, it is clear, how does it gauge against that experience of really knowing? To be able to use that information to know whether what we are getting is real or not. I think that is one way of knowing whether what we know is fantasy or not.

I think also we need to keep pushing ourselves to make that connection at the deepest level that we are able to do it. To figure out what our process is of doing that. How we do that, what steps we take, how we know when we have gotten there. What kinds of cues there are that give us that information. So it becomes a conditioned response that we can do easily with very little fuss and bother. Like when we first start centering: at the beginning we have no idea what we are doing. We have no idea what we are aiming for, and as we do it over and over again, we find it becomes easier because the mechanics of it are not so much of a struggle any more. It's like learning to drive a standard. At the beginning it's all the tasks you have to do first. Then it becomes second nature. The more we are really clear with ourselves about the way that we center, the way that we assess, the way we make that connection with our client, the way that we treat and reassess, how we know the cues we got, and when they've shifted—how we know that. How we know when we are centered and when we have lost it. All of those pieces of information we need to keep revisiting and keep being really clear with ourselves about what our experience is because when we do that we are able to articulate it. We can write it down—I don't mean at the time we are treating, but later—go back to the experience and pick out some of the information. What starts to happen is that it occurs more easily and with less conscious thought and more automatically. The quality of information you get tends to be better and more complete. That certainly is what has happened for me in terms of my own practice. That is what I encourage my students to do as well.

D: Yes, your assessment goes deeper and deeper, as I am sure that you know.

A: Yes, that is right.

D: Does anybody else have more?

A: As I listen to people talk about trust, I think I have a slightly different experience, just in the sense that over the years, I have developed a greater sense of trust, but I notice there is also a bit of the skeptic that still speaks. Both are present. I equally feel more trusting in the process but there is still the skeptic that says, "How do you know that? Is this really real? You need to check it out." So on the one level the trust is deepening so that there are parts of it that I don't question as much and I can know. But some part is still in the back and is still questioning things.

D: Okay, I think we are really beginning to get there. I really do. Now I would like to change things a little. In this second part I would like to show you some slides that were taken by Loren (see page 128–138 photos): a whole series of these two gals doing TT to a man. Stacey and Jan are doing TT to a man by the name of Jim who is ill. I would like to show you each slide, and feel free to call out the kinds of things you see Stacey and Jan doing. Shout out the interpretation of what it is that is going on within them.

After the slides on the TT treatment you will see an example of peer therapeutics. We have been doing peer therapeutics for perhaps twenty years now. Essentially it is this: while you are doing TT, the patient gets interested in what you are doing, or admires you as healer so that they get interested in the process itself. Once they have that interest and we think it will be to the healee's benefit, we teach them an introduction to TT and when they are good at it and also safe, we get them to do TT on a patient. This patient, under the best of circumstances, has the same problem as the new healer has, whether it is a disease process or trauma or whatever. The healee (now healer) begins to work with the new patient, with a kind of identification . . . a memory . . . of their own previous experience. And out of that healing interaction there have been wonderful insights and breakthroughs in terms of what that illness meant to that person now playing the role of healer—what the reason was for the illness, for instance—and so we encourage the healees to learn TT and then do TT as I noted in this engagement of peer therapeutics that fosters insight so very well.

Okay, let's begin with the slides. That's Stacey on the right and Jan on the left. What are the kinds of things you are seeing there?

A: They are all centering.

D: They are all centering . . . yes, more . . .

A: They are cocooning Jim while they do it.

D: Yes, more . . . . What are the things you see? There's centering, true . . .

A: They are creating a field.

D: Yes, that's fine . . . creating a field . . . . What else do you see?

A: There's an openness that they are creating—to each other and to the client.

D: Okay, what else do you see?

A: They are sensing each other's field.

D: Okay, guess what my next statement is going to be?

A: More, more . . . !

D: Yes. More. There's more . . .

A: They're grounding.

D: More . . .

A: They are welcoming the available energy from above.

D: Okay, that's good. More . . .

A: I think they are going internal.

D: They are going eternal? (Laughter.)

A: They don't appear to be talking.

D: Okay, more, much more . . .

A: They are connected to the evolving patterning of energy.

D: Hey, there you have it. They are connected to the developing pattern of energy. Now take a look at Jim's back. What do you see there? It's tension. You see? Look at the neck.

A: Because they are centering behind Jim's back. He doesn't see what is going on.

D: That's very true. But you really can't miss that, can you? Look at the tension in his back. He is almost like a stick. Watch what begins to happen to his body as the TT progresses. Is there anything else we haven't looked at? Look at the intent look in the faces.

A: The eye contact with each other.

D: Yes, very fine.

A: The eye contact—they look like they are connected.

D: Tracey, can we have the next slide? This is only slightly different, but see how they are allowing their hands to descend through the field. You can see the tension real well in Jim. See that neck. The head is bent forward. He is a pretty big man. What is happening now?

A: The shoulders are dropping.

D: Okay.

A: They are moving in and beginning the assessment.

D: Is there anything else you want to say on that one?

A: The focus has actually shifted. Before the focus was on openness, on setting up communication, and connection. Now it is the work of the assessment has begun. They are beginning to bring the information to themselves through that open connection.

D: Okay, that's good. Very observant. On to the next one. Can you see the intent now?

A: Yes, the concentration, the focus, the listening. The shoulders are dropping.

D: You can still see tension in the neck there in the sternocleidomastoid. I remember that. . . . (Laughter.) Does it tell you anything at all about what is going on besides the fact that they are using it to keep an alignment with where they are going?

A: There's a great respect. Yes, there is a reverence. And very gentle, too. And his face isn't tuning out.

D: I think that what happened with this particular patient was that Stacey was particularly attracted or interested in his lymphatic system. She is getting a lot of feedback from it. I don't know if you have noticed it; her hands have changed position.

A: One thing that I noticed is that at the first part when Stacey was in the front, her knees were locked in tension.

D: Stacey. Yes, she is tall.

A: You see his face. He's changed. Look at the way his neck is, too. His eyes have changed, too.

D: That was an interesting shot of Jan, too, wasn't it? Caught in mid-stream. This is when she begins to . . . well, she is not going for his heart. She begins to . . . . Look at the look on her face. She got interested in the lymphatic system.

A: Yes, yes . . .

D: She is going after it. . . . She is very sensitive. They are both very sensitive, but Stacey is unusually sensitive. See the way she is watching Jan.

A: His spine has straightened up. It was curved forward but now he has straightened up.

D: Yes, that's right.

A. Oh . . . ah . . .

D: A different looking man, eh?

A: Oh . . . Ah . . .

D: That is a shot of some joke he was making. You can see the good relationship. My God, he does look like a different man now. Doesn't he?

A: Younger . . .

D: He is just a . . . they are sort of friends, that's all. . . . Now, this slide begins to demonstrate the peer therapeutics. He is working with a TT therapist named Denise, and Carolyn is the person they are working on.

A: How long is this after his treatment? Another day?

D: Yes, another day. . . . I think his technique is good. He seems very much at home. . . . He has pretty good technique, I think. Okay, so we have come to the end of the slides. Has anybody got anything to say in terms of the slides?

A: He seems so happy to be able to help, you know? The shift in him from being helped to wanting to help. As far as demeanor and the way he is even touching or interacting. It is really very beautiful. You can just sense the compassion for the other person. It enabled him to do that. Allowed him to be able to do that. You can see the change in him.

D: Anybody has anything more to say? Particularly those who haven't said anything. You are sitting there going to sleep on me! (Laughter.)

A: You can really feel what is going on.
D: Yes, the energy of it is almost palpable.
A: I was really struck by how much you could tell from just looking at the pictures.
D: Is there anybody else that has anything to say about this?
A: It just surprises me how similar the process is. Those people from way over on the East Coast and us from out on the West Coast . . . I have never met these people before and they get together and we know just what they are going to do. . . . (Laughter.)
D: At least you're Canadian and you speak English! (Laughter.) Okay, well, as you realize, the whole reason for tonight was to give you a mirror on yourselves. You are really beautiful. This is really what is happening. I think some of the expressions tonight . . . you really hit it on the nail. I hope it gives you a lot to think about. It is hard to get a gauge of what you yourself are doing, but you are perfectly right. There are more similarities among us than differences. And it is beautiful. I have oftentimes thought of TT, as some of you know, as a dance, as a pavane. Very stately and elegant. Is there anything else?
A: It is always a thrill to work with somebody a couple days in a row like our clients that we worked with yesterday and today. To see the shift in energy as far as knowingness, as far as connecting with the others in the group, as far as words that came out of her mouth to describe how she is feeling, her own description. But besides that a change in her aspect. I think it was quite clear to the others in the group that there was definitely a shift, and then yesterday our client came just in time for the treatment, went home immediately after she finished her rest period. Today I believe she would have stayed longer except that the smoke came out of the stove in the teepee and affected her breathing. She cannot do that so she said, "I have to move." But she at least stayed for the discussion afterward. She said, "I would like to stay but I have to move." But it was amazing. It was really a thrill to see the struggle. To see several of them getting

together later. She also indicated she would like to get involved with a patient group at some point before the end of the week. She just wanted to give herself a couple of days to get her feet back under her again. And I really think she will.

A: A lot of the people in our group came from physical therapy and from Roseanne's community health group. It's great for me, as the therapist, to have this support because I work alone. This is so great to have your professional feedback about our patients in PT and community health. You don't realize how it opens the door to the community. It's great. I don't have to be the one to say everything. I let other people say something. It's in a different voice. Now when the patients come in they will have a whole different point of view. This is a vital link to the immediate community. There has not been that exchange readily. Just look at all the beauty you are bringing the patients. So we have a wonderful opportunity and variety. But the really hard client hasn't come yet. I want her to come so I am hoping she will tomorrow.

D: That will be nice.

A: I think it is a really important thing that you brought up. We need to be aware of that in our own communities. That [. . .] it behooves us to also have a friend who comes in as an outside expert. We stand back and let them talk . . . because it adds such credibility. That's what it is. Credibility. And it reinforces the intimate trust that is already being built up. And they are people from different walks of life.

A: We all know what the definition of an expert is. It is someone who is not from where you are and is not a member of your family. (Laughter.)

D: I have enjoyed this evening. I think we have all learned a great deal. Thank you very much.

Come to the edge, s/he said.
They said: We are afraid.
Come to the edge, s/he said.
They came.
S/he pushed them . . . and they flew.

—Guillaume Apollinaire

APPENDIX I: RANDOM STUDENT PROFILES

Profile I

Note: In the randomly selected class of thirty-four students in this survey, some work in more than one setting, and some do Therapeutic Touch with persons who have several conditions. Therefore, this is a broad-spectrum survey that depicts their major interests and functions re Therapeutic Touch (TT).

A) Sites where students are doing Therapeutic Touch

Sixteen students work in hospital settings. Three have developed in-house TT practice programs, two supervise other health personnel while they do Therapeutic Touch on patients, and three teach TT in in-service programs. Two students do TT to persons who are clients in outpatient departments, seven do TT in mental health settings, five do TT in pre- and postoperative settings, three work in intensive care units, another three students are on rehabilitation services, and five do TT in hospice facilities. In addition to working in institutional settings, five have an independent TT practice, and one person teaches TT to staff at hospitals as part of her independent practice. One person works with a community self care group, two work within other community health agencies, and three students work with families.

B) Major conditions treated with Therapeutic Touch

Five persons do TT to pre- and postoperative patients, two work with patients whose problems are due to the aging process, seven use TT with other patients with chronic conditions, six do TT with oncological patients, and three use TT particularly with persons who are dying. Two persons do TT during labor and delivery, one works with people who have substance-related addictions, one person does TT to patients in the emergency room, and two use TT to accelerate wound healing.

Four persons combine TT with massage sessions, four do TT for people in high stress situations, and four use TT specifically to elicit a relaxation response in people under unusual duress. Three use TT for patients with arthritis; in addition, five persons use TT for people with other severe pain. Two persons use TT for fibromyalgia and chronic fatigue, three do TT for sports injuries and occupation-related injuries, three work with patients with back pain and other musculoskeletal injuries. Three people use TT with persons with allergies, autoimmune diseases, and upper respiratory infections.

Several persons use TT with people with emotional problems: three use it for persons with depression, three use TT with persons who are highly anxious, another three use it for agitation in persons with Alzheimer's disease, one person uses it for trauma release, and another does TT treatments with persons suffering from sustained childhood trauma.

Profile II

Note: Of the sixty-two TT therapists in this group, some students work in more than one setting, and some do Therapeutic Touch with persons who have several vital-energy system imbalances. Therefore, this is presented as a random, broad-spectrum survey to indicate this group's particular interests and functions re Therapeutic Touch, and is not a formal study.

A) Sites where Therapeutic Touch is being done

Thirty-seven TT therapists work in hospitals, six do TT in hospices, eleven work in community health agencies, thirteen do TT in chronic care facilities, senior centers, etc. Fourteen work with patients in acute medical care centers, twenty-two do TT on various surgical services, eleven work in either emergency services, trauma rooms, or intensive care units. Twelve work in psychiatric facilities, ten in ambulatory clinics, and three in correctional care health services. Thirty-four TT therapists have private practices, and nine teach TT in university settings. Fifteen of the above-noted facilities have approved procedures and policies for Therapeutic Touch practice in-house.

B) Major conditions treated with Therapeutic Touch

Thirty-six TT therapists work with persons with various cancers, nine work with persons with AIDS, thirty-two do TT with neurological patients. Thirty-two others work with patients with chronic fatigue and/or fibromyalgia, fourteen work with persons having Alzheimer's disease, and thirty-nine do TT with persons in pain. Six persons do TT with pregnant couples during the birthing process, twenty-five work with children, and six do TT in high-wellness centers.

APPENDIX II: THERAPEUTIC TOUCH CONTACTS

Long-term courses in Therapeutic Touch

Pumpkin Hollow Foundation
1184 Route 11
Craryville, NY 12521
phone: 518-325-3583
fax: 518-325-5633
email: pumpkin@taconic.net

Therapeutic Touch Network of Ontario
123 Queen Street West
Brampton, ON, Canada L6Y1M3
phone: 905-454-2688
fax: 905-453-3747
email: marydale@idirect.com

Orcas Island Foundation
360 Indralaya Road
Eastsound, WA 98245
Phone: 360-376-4526
Fax: 360-376-5977
Email: oif@rockisland.com

APPENDIX III: SPECIALTIES

The Official Organization of Therapeutic Touch

Nurse-Healers Professional Associates International
3760 South Highland Drive, Suite #429
Salt Lake City, Utah 84106
801-273-3399 phone
901-273-3352 fax
NH-PAI@therapeutic-touch.org

Specialties recognized among the members of NH-PAI

Acupuncture
Administration/Management
Aromatherapy
Bach Flower Remedies
Bioenergetics
Biofeedback
Cardiopulmonary
Chinese Medicine
Color Therapy
Community/Public Health
Consultation/Liaison
Counseling/Psychotherapy
Craniosacral Therapy
Crystal Therapy
Dance/Movement Therapy
Education
Emergency Care
Feldenkrais Method
Gerontology
Healing Touch

Herbology
Home Care
Homeopathy
Hospice
Imagery/Visualization
Iridology
Kinesiology
Massage Therapy
Maternal & Child Health
Medical-Surgical
Meditation
Midwifery
Music
Naturopathy
Neurolinguistic Programming
Neurology
Neuromuscular Therapy
Nurse Practitioner
Nutrition
Orthopedics
Outpatient/Primary Care
Pediatrics
Polarity
Reflexology
Health Rehabilitation
Reiki
Renal Care
Research
Rolfing
Rosen Method
Shiatsu
SOMA
Staff Development/Education
Stress Management
Tai Chi
Therapeutic Touch for Health
Trigger Point Myotherapy
Trager Psychophysical
Integration
Touch Yoga
Women's Health

APPENDIX IV: KRIEGER'S THEORY OF NEEDS RE TEACHING

Occasionally, it's useful to bring humor into your teaching. I share this theory with my students for laughs.

Teaching is an Amalgam of:

- Masochism (A Need-to-Prove)
- Knowledge (A Need-to-Know)
- Altruism (A Need-to-Share)
- Showmanship (A Need-to-Brag!)

APPENDIX V: TEN COMMANDMENTS ON GIVING A WORKSHOP ON HEALING

Thou Shalt:

1. Have something significant to say.
2. Be clear in understanding your own motivation to teach this content to others.
3. Assure conditions that will constructively support the transmission of information. The milieu in which the workshop will be held, the theme, the sponsorship, and the relationship between you and the sponsor(s) should be clearly based on the freedoms you need to develop the kind of workshop of which you can be proud.
4. Be sure your objectives for the spectrum of experience you are offering are clearly stated, and be discriminating about the participants; they should be able to bring with them the background and the tools that will assure them a significant learning experience.
5. Judiciously use teaching aids, such as audiovisual material, workshop outlines, and selective bibliographies that will help participants in their learning process.
6. Intelligently limit the scope of the presentation to meet the special needs of the learners.

7. Request written, personal objectives of the workshop from each participant to mold the workshop content so that it is individually relevant.
8. Underline the word "work" in your workshop, and involve the participants meaningfully and experientially in workshop content.
9. At the end of each session, briefly summarize the principles underlying the workshop experience thus far.
10. At the end of the workshop, request a written evaluation from each participant, and use the comments to refine your teaching techniques for your next workshop.

APPENDIX VI: QUESTIONS THE THERAPEUTIC TOUCH TEACHER NEEDS TO ASK

Note: As the student begins to explore the TT process in depth, her progress is marked by deep-seated, often significant changes in the perception of her life's events and the growing recognition of how she can change them. Being of a positive, beneficient nature, the influence of TT in one's life is most often heartening, enlightening, and exhilarating. However, the TT teacher's responsibility naturally extends to include the role of mentor to her students. As mentor, the TT teacher is cognizant of the growing, conscious though complex relationship the student may develop with her inner self; thus, consideration for her students' physical and psychological well-being extends beyond the simple teacher-student affiliation.

Because of the complexity of this personal growth, the thoughtful teacher guides her students beyond the ordinary theoretical context and practice of Therapeutic Touch to include the increasingly conscious dialogue the student may engage in with her inner self. To assure a healthy progression of the student's learning style, below are several questions that the concerned mentor might ask her students as they progress into the advanced stages of TT practice.

- It is compassion for those who are ill or in crisis that most often urges a person to help or heal those in need.

Compassion is a precious human attribute and a powerful machine in the forging of the healing process. Its nature is such that it fosters in the TT therapist the courage to help or heal people without being attached to the results for personal benefit. To do this, one must learn to act solely for the betterment of the healee. Is your student able to differentiate clearly compassion from mere sentimentality, sympathy, or empathy?

- The act of centering the consciousness and maintaining that state of consciousness throughout the TT interaction acts as the point of entry to the TT process. For the TT therapist, what, or who, is at center during this time?
- Several basic assumptions form the foundation for the Therapeutic Touch process. How does the student put these assumptions to work in her daily life?
- One such assumption concerns a recognition that order, rather than randomness or chaos, underlies the natural laws of the universe. How has this concept of order changed the student's worldview?
- The human energy field encompasses the potentialities of our functions, emotions, concepts, and aspirations. What control does the TT therapist have over her own field faculties?
- The vital-energy field is particularly concerned with the physiological aspects of one's being. How have these functions changed as the TT therapist's healing work has progressed?
- Therapeutic Touch is said to be a process. How does the TT therapist know this to be true (i.e., that it is not a mere stimulus-response reflex)?
- "Listening" for cues is the key to the TT assessment. How does the mindful TT therapist test for signs of imbalance in the rhythmicity of the healee's vital-energy and psychody-

namic fields and the consequent change in patterning in these fields?

- Through the years, many TT therapists took to abbreviating the term for the inner self as IS, for it IS the enduring part of oneself. They recently personalized the idea as ISSIE (pronounced Izzie). In considering the process of the TT therapist's awakening to the presence of her inner self, of what nature is the establishment of this relationship with ISSIE?
- As the TT therapist progresses in her healing work, she continually works with her chakra complex, particularly the hand chakras. Can the TT therapist make these chakras work on command?
- With the increased conscious use of her chakra complex, the TT therapist experiences a deepening of her various levels of consciousness. How does she check on their validity?
- With the continued conscious use of intentionality during the rebalancing phase of Therapeutic Touch, the TT therapist may experience true visualizations of the dynamisms of this process. Often this is a transverbal experience. However, by using appropriate analogues, how does the TT therapist describe this experience?
- With advanced experiences in the TT process, the therapist may begin to realize the transpersonal nature of the TT interaction with the healee. How has the TT therapist explained to herself this level of discourse?
- The advanced TT therapist may be able to transfer the lessons she has learned through her healing interactions so well that she may transcend personal circumstances in her life. What is the leading edge of that change, and how does she incorporate into her life what she has learned from the frequent use of the TT process?

APPENDIX VII: WHAT IS A MENTOR?

The concept of Mentorship goes back to the earliest teaching. The Mentor as archetype, e.g., the Wise Woman, the Sage, the Teacher, most often was configured as the guiding spirit of the Age, as one who entered into the individual's personal myth and activated it into reality.

As guardian spirit, the Mentor was one who watched over the welfare of the people—the one to whom an individual prayed for favors, for instance. These guardians often took the form of birds, animals, and nature spirits.

In the Celtic tradition, the Mentor was called *anam cara*, translated as "friend of the soul," spiritual guide, or teacher in a context that was associated with personal transformation.

In the ancient Greek tradition, the Mentor was conceived as "the power invisible" who provided access to the Divine.

More recently, the concept has been clothed in a Jungian framework that implies that mentorship reflects and draws out of each of us our own beauty, strength, and wisdom. Today's mentor is therefore conceived as a wise, experienced, and trusted advisor, someone who is sensible, prudent, knowing, understanding, and authoritative.

What You Can Expect from Your Mentor

As noted, the mentor is regarded as an individual's teacher, guide, model, or, in the current paradigm, as a personal trainer, one might say. As mentoree, you may be receiving many favors of wisdom or connections to high places and Very Important Persons. You will be dealing with the ethics of the "real" (i.e., real tough, real competitive) world, the dynamics of power in the realm of professional and business connections, and the intricacies of "how to play the game." You'll also be dealing with the nuances of the development of your personal style, and, very importantly, with your individual empowerment.

You will learn the essential tricks of the trade, e.g., how to organize your expertise, and how to demonstrate it effectively. It is fair game for your mentors to expect something in return. Therefore, as mentoree you are essentially an apprentice: you get to vacuum the floor and do the dishes after the party is over. You will gain respect for and feel loyalty to your mentor for the many kindnesses, but don't be coerced into beliefs or actions that do not appeal to your own inner voice, your conscience. Think through for yourself issues of concern, and then take personal responsibility for them.

So, what to look for in a mentor/mentoree relationship? Ask yourself:

- As a Model:
  - Do I want to identify with her?
  - Does she have ego strength, and yet an inner seeking directedness?

- As a Teacher:
  - Can she teach me what I need to know?
  - Can she inspire me to learn?
  - Can she guide me in the self-search, the personal quest I have to take to be a healer?

- As a Protector:
  - Will she protect my vulnerability?
  - Will she protect my aspirations?
  - Will she protect my naivety and innocence?
  - Will she protect my professional reputation?

- As an Advocate:
  - Will she act in my best interests?
  - Will she truly represent my best qualities?
  - Has she the insight to understand me?
  - Does she appreciate my potential?

Your mentor will help you to mold the near future of your life. Choose wisely for the long haul—then work like hell and prove that you are worthy of her trust, her time, her wisdom, her creativity. And then, when you're ready to stand on your own—go for it!

For in time you too can prove worthy of becoming a mentor, and those mentorees will have their mentorees, and so on and so on, ad infinitum.

APPENDIX VIII: VISUALIZATION DURING THERAPEUTIC TOUCH

Equipment

Pen, paper, and/or tape recorder

Information

- Deeply center as you do the assessment on a healee during a Therapeutic Touch session.
- As you pick up cues and intentionality wells up about how you will rebalance the healee's vital-energy field, you may have a fleeting visualization of the rebalancing process, an insight into the origin or meaning of the healee's illness, etc.
- When/if your cues constellate into a stable, clearer visualization, go "out" to the object of your perception by gently and sensitively "stretching the edges" of your vital-energy field, meanwhile being sensitively aware of the pulsations of the healee's vital-energy field.
- Try to "feel" the substance of the healee's vital-energy field with your own vital-energy field; try to "touch" it mentally, picking up whatever information impresses you.

- Record your impressions, using prose, poetry, haiku, movement, vocalization or the humming of your feeling tone of the experience, or other means of your own choosing.
- Read or play back the expression of your experience and try to determine the meaning it has for you. Consider your interpretation in the perspective of what you know about the healee from all other means.

APPENDIX IX: SELF-EVALUATION OF THERAPEUTIC TOUCH SCALE (SETTS)

A Self-Evaluation of Therapeutic Touch

The Self Evaluation of Therapeutic Touch Scale (SETTS)[1] was developed at New York University in 1983 by Dolores Krieger, Ph.D., R.N. and Patricia Winstead-Fry, Ph.D., R.N. An in-depth statistical analysis, which replicated its findings, was done by Cecelia K. Ferguson, Ph.D., R.N. as her doctoral dissertation, "Subjective Experience of Therapeutic Touch: Psychometric Examination of an Instrument."[2] The purpose of SETTS is to distinguish persons who are experienced in the practice of Therapeutic Touch from those who are inexperienced.

The internal consistency reliability of SETTS was measured by Cronbach's alpha in both studies; in the Krieger and Winstead-Fry original study the Cronbach's alpha was 0.98, and in Ferguson's psychometric analysis the coefficient was 0.97, thereby confirming the Krieger and Winstead-Fry findings. Validity of the instrument was also acknowledged. SETTS was found to have the ability to differentiate experienced from inexperienced practitioners of Therapeutic Touch ($P > .01$),[3] and also to distinguish clearly between inexperienced nurse practitioners of Therapeutic Touch and those nurses who did not practice Therapeutic Touch ($P > .01$).

Multiple linear regression indicated that two factors contributed most to a high score of SETTS:

1. The frequency with which clients reported improvement in their symptoms; i.e., the greater the practitioner's expertise, the higher her score
2. The number of people treated by the practitioner, so that the greater the amount of experience the Therapeutic Touch therapist had, the higher her score on SETTS

The highest possible score on SETTS is 272. The means (average) of the experienced Therapeutic Touch therapist was 177; the means of the inexperienced Therapeutic Touch therapist was 137.

In her replicative study, Ferguson concluded that the persons who are experienced in Therapeutic Touch describe themselves differently than do persons who are not, particularly in the areas of nurturance and creative personality.

After the student has had several weeks' experience in the practice of Therapeutic Touch, he or she should fill out a copy of SETTS. This self-evaluation practicum can be done either alone or in small groups of two, four, or six people (the most effective size seems to be four). One person in each group should volunteer to take notes.

- The persons in each small group read their statements to each other, each supporting her self-evaluation by describing recent appropriate TT experiences she has had.
- After completing SETTS, consider the following questions for discussion in small groups:
  - What came up during the relating of your group members' experiences that was similar to your experiences?
  - What in your group members' accounts of their experiences surprised you?
  - What did you learn from exchanging these experiences?
  - In looking back at your own experiences with the TT process, what would you do differently?

Note to students: Although valid research findings are relevant only within a particular context, SETTS is reproduced below as an informal gauge of the TT therapist's interior or subjective experience with the TT process. As you proceed with this study of your own increasing expertise, SETTS will help clarify for you the nature of the act of centering, which is so central to the Therapeutic Touch process and to the powers of mindfulness that it fosters. There is a rough relationship of statements in SETTS to the following variables:

A. Centered state of consciousness:
   1, 2, 10–14, 17–19, 21–31, 35, 37–40, 45–50, 52, 54–60, 64–68.

B. Therapeutic Touch assessment
   3–8, 32–36, 41–44, 51.

C. Therapeutic Touch rebalancing
   9, 15, 16, 20, 53, 61–63

The following items reflect experiences that Therapeutic Touch practitioners have had while performing Therapeutic Touch on healees. For each of the items, mark the frequency with which the experience occurs to you while you are engaged in the process of Therapeutic Touch using the following scale:

0 - not at all
1 - once in a while
2 - frequently
3 - almost always
4 - all the time

Rating Your Experience of Therapeutic Touch

__1. My heart and respiration rates feel slower.
__2. My breathing becomes slower and deeper.

__3. I feel sensations of heat and cold in my hands.
__4. I feel tingling sensations in my hands.
__5. I feel pressure in my hands.
__6. I feel electric shock sensations in my hands.
__7. I feel energy pulsations in my hands.
__8. I have the feeling that my hands are being spontaneously drawn to a particular area in the healee's field.
__9. I feel heat coming from my hands.
__10. I seem to be able to maintain uncomfortable postures much longer than usual.
__11. I seem to stand or kneel straighter than usual.
__12. My body movements become subtle, soft, and flowing.
__13. I become very sensitive to how I move my body and whether I am in an awkward or stressful position.
__14. My movements feel slow, steady, smooth, and alert.
__15. I feel energy moving through me and out of my hands.
__16. Energy flows more freely in my body.
__17. I get a sense of stillness and balance in my body, mind, and emotions.
__18. My body feels in harmony and seems to be an instrument through which energy flows.
__19. My body feels quiet, calm, and relaxed.
__20. I feel energy flowing rhythmically and evenly within my body.
__21. I feel physically balanced, lined up, or integrated.
__22. I feel as though all the parts of my body are working in unison.
__23. I have a sense of physical and psychological attunement.
__24. All my senses are heightened and sharpened.
__25. I feel very close to the person I am healing.
__26. I feel personal love for the healee, regardless of whether I liked the person before or not.
__27. I feel loving and accepting toward myself and the healee.

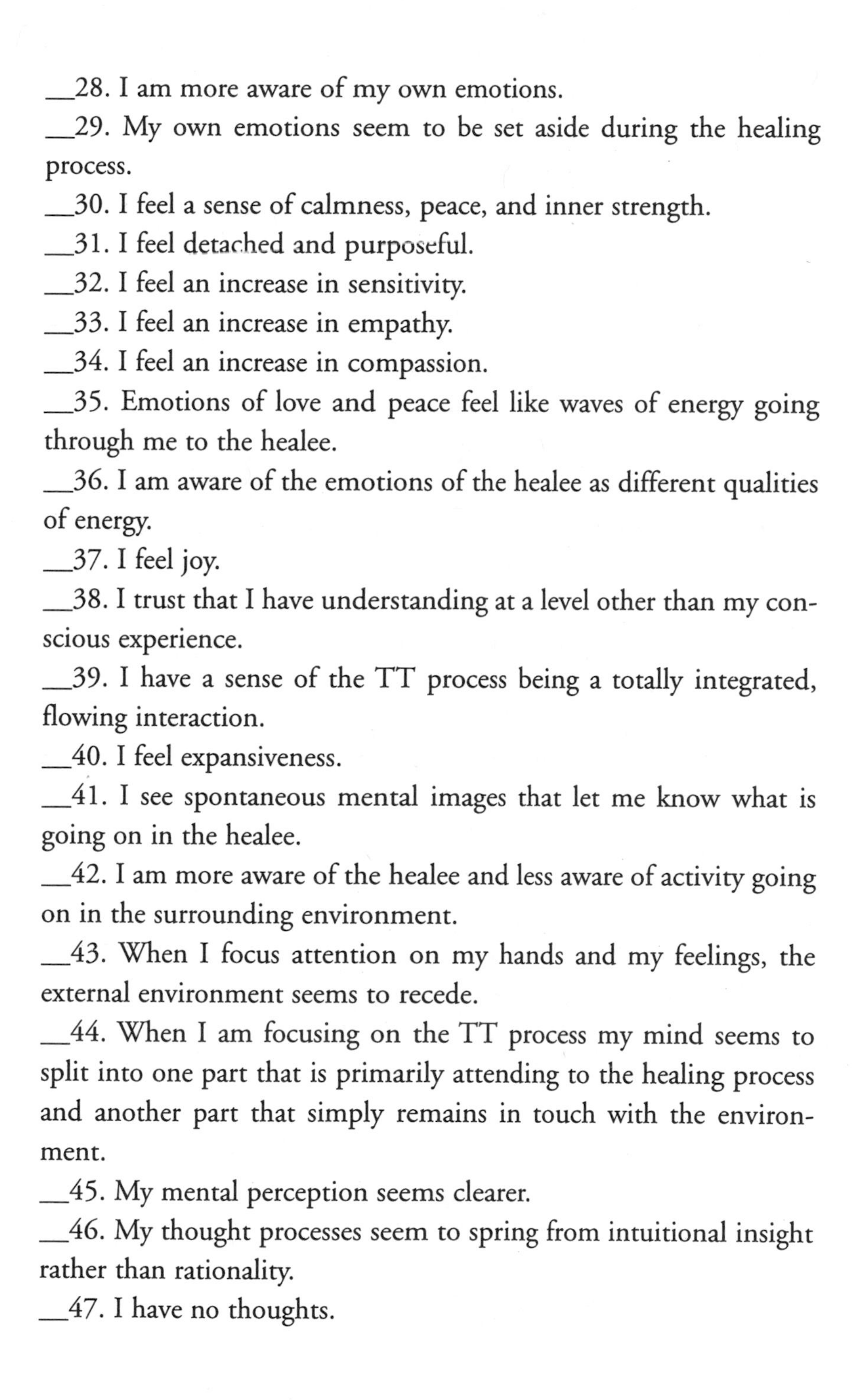

__28. I am more aware of my own emotions.
__29. My own emotions seem to be set aside during the healing process.
__30. I feel a sense of calmness, peace, and inner strength.
__31. I feel detached and purposeful.
__32. I feel an increase in sensitivity.
__33. I feel an increase in empathy.
__34. I feel an increase in compassion.
__35. Emotions of love and peace feel like waves of energy going through me to the healee.
__36. I am aware of the emotions of the healee as different qualities of energy.
__37. I feel joy.
__38. I trust that I have understanding at a level other than my conscious experience.
__39. I have a sense of the TT process being a totally integrated, flowing interaction.
__40. I feel expansiveness.
__41. I see spontaneous mental images that let me know what is going on in the healee.
__42. I am more aware of the healee and less aware of activity going on in the surrounding environment.
__43. When I focus attention on my hands and my feelings, the external environment seems to recede.
__44. When I am focusing on the TT process my mind seems to split into one part that is primarily attending to the healing process and another part that simply remains in touch with the environment.
__45. My mental perception seems clearer.
__46. My thought processes seem to spring from intuitional insight rather than rationality.
__47. I have no thoughts.

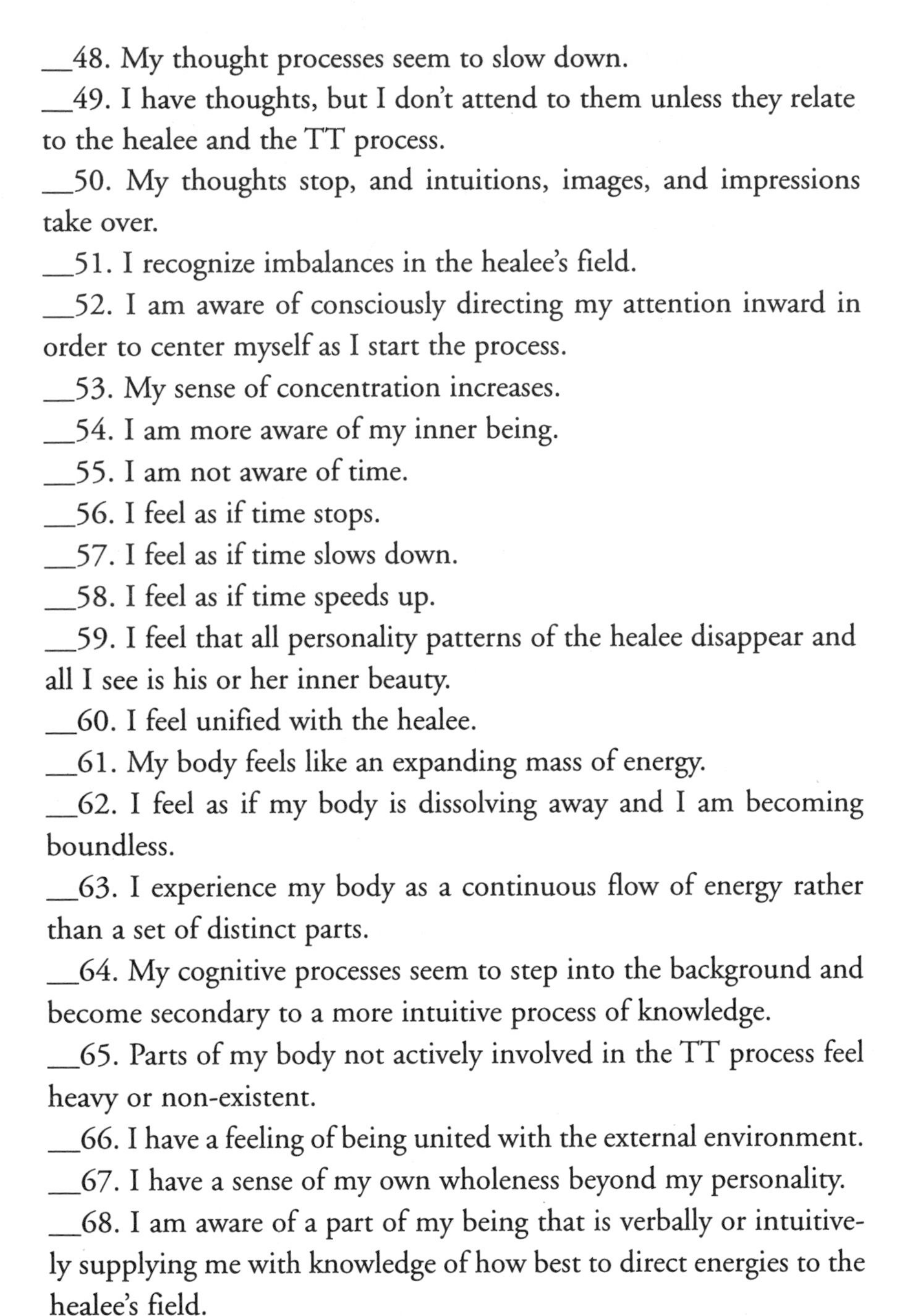

__48. My thought processes seem to slow down.
__49. I have thoughts, but I don't attend to them unless they relate to the healee and the TT process.
__50. My thoughts stop, and intuitions, images, and impressions take over.
__51. I recognize imbalances in the healee's field.
__52. I am aware of consciously directing my attention inward in order to center myself as I start the process.
__53. My sense of concentration increases.
__54. I am more aware of my inner being.
__55. I am not aware of time.
__56. I feel as if time stops.
__57. I feel as if time slows down.
__58. I feel as if time speeds up.
__59. I feel that all personality patterns of the healee disappear and all I see is his or her inner beauty.
__60. I feel unified with the healee.
__61. My body feels like an expanding mass of energy.
__62. I feel as if my body is dissolving away and I am becoming boundless.
__63. I experience my body as a continuous flow of energy rather than a set of distinct parts.
__64. My cognitive processes seem to step into the background and become secondary to a more intuitive process of knowledge.
__65. Parts of my body not actively involved in the TT process feel heavy or non-existent.
__66. I have a feeling of being united with the external environment.
__67. I have a sense of my own wholeness beyond my personality.
__68. I am aware of a part of my being that is verbally or intuitively supplying me with knowledge of how best to direct energies to the healee's field.

Developed 1983 by Dolores Krieger and Patricia Winstead-Fry, Ph.D., R.N.

APPENDIX X: SOURCES OF INFORMATION ABOUT THERAPEUTIC TOUCH

Information on Therapeutic Touch may come from a variety of sources. It is usually in the form of factual data from case histories, laboratory findings, demographic data, physiological and psychological test findings, or other types of tests and medical imageries. Another source of information is research findings. A reference list can be accessed through Nurse Healers-Professional Associates International. Another resource for TT-related research is MediCom Worldwide (www.medicaled.com). For doctoral dissertations, see the annual *Dissertation Abstracts*, which can be consulted at any university library, major public libraries, and the libraries of some professional societies. Most postdoctoral research on the TT process is reported in professional journals and other professional periodicals, whereas clinical studies are usually published in specialty journals, or a variety of other sources of applied research reports.

If research has been funded by a grant, a report of the research findings eventually is made to the institution funding the grant. These reports are usually available to other researchers, teachers, or graduate students. A wide gradation of studies on Therapeutic Touch can be found in various avenues of the media, from the serious to the sensational, and the reader is cautioned to use discrimination in their use and discretion in interpreting them.

NOTES

Chapter I. What Is Healing? A Brief History

1. Inglis, Brian. *The Diseases of Civilization.* London: Hodder and Staughton, 1981, pp. 284–288.
2. Worrall, A. A. and O. N. Worrall. *The Gift of Healing.* New York: Harper and Row, 1965.
3. Krieger, Dolores. *Foundations for Holistic Health Nursing Practices: The Renaissance Nurse.* Philadelphia: Lippincott Williams & Wilkins, 1981, pp. 140–147.
_______. *Living the Therapeutic Touch: Healing as a Lifestyle.* New York: Dodd, Mead & Company, 1987, pp. 12–15, 167–187.
4. Krieger, Dolores. *Therapeutic Touch Inner Workbook: Ventures in Transpersonal Healing.* Santa Fe: Bear & Co., 1996, pp. 176–186.
5. Krieger, Dolores. *Accepting Your Power to Heal: The Personal Practice of Therapeutic Touch.* Santa Fe: Bear & Co., 1993.
6. Abraham Maslow. *Motivation and Personality.* New York: Harper, 1954.

Chapter II. Concepts and Constructs Essential to the TT Process

Chapter III. Clusters of Concepts and Constructs Relevant to Therapeutic Touch

1. Newton, Isaac. *The Opticks.* 1704.
2. Heisenberg, Werner. *Across the Frontiers.* New York: Harper & Row, 1974.

Chapter IV. The Universal Healing Field

1. Einstein, Albert. *Ideas & Opinions.* New York: Crown, 1954.
2. Pagels, Heinz R. *The Cosmic Code: Quantum Physics as the Language of Nature.* New York: Bantam Books, 1983, p. 241.
3. Kunz, F. L. "The Reality of the Non-Material," *Main Currents in Modern Thought,* Vol. 20, No. 2, Nov.-Dec. 1963, pp. 34, 37.

Chapter V. The Human-Energy Field

1. Burr, H. S. and C. T. Land. "Electrical Characteristics of Living Systems," *Yale Journal of Biology and Medicine,* 8:31–35, 1935.
______, and F. S. C. Northrup. "The Electro-dynamic theory of Life," *Main Currents in Modern Thought,* Vol. 19, No. 1, Sept.-Oct. 1963, pp. 4–10.
Ravitz, Leonard J. "Studies of Man in the Life Field," *Main Currents in Modern Thought,* Vol. 19, No. 1, Sept.-Oct. 1962, p. 14.
2. White, John. *Frontiers of Consciousness: The Meeting Ground Between Inner and Outer Reality.* New York: Julian Press, 1974, p. 557.

Chapter VI. The Nature of Consciousness

1. Penfield, Wilder. *The Mystery of the Mind: A Critical Study of Consciousness and the Human Brain.* Princeton, NJ: Princeton University Press, 1975.
2. Teilhard de Chardin, Pierre. *The Phenomenon of Man.* New York: Harper & Row, 1975.
3. Tart, Charles T. *Altered States of Consciousness.* New York: John Wiley & Sons, 1969.
4. Cannon, Walter B. *The Wisdom of the Body.* New York: W. W. Norton & Company, 1942.
5. Selye, Hans. *The Stress of Life.* New York: McGraw-Hill, 1956.
6. Harmon, Willis. *Global Mind Change.* San Francisco: Berrett-Koehler Publishers, Inc., 1998, p. xvii.
7. Maslow, Abraham. *Toward a Psychology of Being.* New York: Van Nostrand, 1982.
_______. *The Farther Reaches of Human Nature.* New York: Penguin, 1993.

8. Frankl, Viktor. *Man's Search for Meaning.* New York: Washington Square Press, 1985.
9. Assagioli, Robert. *Psychosynthesis.* New York: Viking Press, 1971.
10. Ferguson, Marilyn. *The Aquarian Conspiracy: Personal and Social Transformation in Our Time,* new ed. New York: G.P. Putnam's Sons, 1987.
11. Naisbitt, John, and Patricia Aburdene. *Megatrends.* New York: William Morrow, 1990, p. 241.
12. Swami Rama. "Energy of Consciousness in the Human Personality." In *Metaphors of Consciousness,* edited by Ronald S. Valle and Rolf von Eckarsberg. New York: Plenum Press, 1981.

Chapter VII. The Chakra Complex: Source of Consciousness

1. Jung, Carl G. "Psychological Commentary on Kundalini Yoga. Lecture IV." In *Spring 1976: An Annual of Archetypal Psychology and Jungian Thought.* New York: Spring Publications, 1976, pp. 21, 27.
2. Karagulla, Shafica, and Dora Van Gelder Kunz. *The Chakras and the Human Energy Fields.* Wheaton, IL: Quest Books, 1989.
3. Avalon, Arthur. *The Serpent Power,* 7th ed. Madras: Ganesh & Co., 1964, p. 109.
4. Ibid.
5. Lama A. Govinda. *Foundations of Tibetan Mysticism.* London: Rider, 1969, p. 140.
6. Ibid, p. 175.

Summary: The Conceptual Frame of Reference for Therapeutic Touch

1. Krieger, Dolores. "High order emergence of the self during therapeutic touch." In *Spiritual Aspects of the Healing Arts,* edited by Dora Kunz. Wheaton, IL: Theosophical Publishing House; 1986, pp. 262–271.
2. Krieger, Dolores. *Therapeutic Touch Inner Workbook: Ventures in Transpersonal Healing.* Santa Fe: Bear & Co., 1996, p. 34.
3. Sperry, R. W. "The great cerebral commissure." *Scientific American* 210 (1964): 42–52.
4. Lewis, Jefferson. *Something Hidden: A Biography of Wilder Penfield.* Garden City, New York: Doubleday, 1981.

Chapter VIII. The Power of Compassion: A Potent Link to the Inner Self

1. Eliade, Mircea. *Shamanism.* Princeton: Princeton University Press (Bollingen series LXXVI), 1972.

Chapter IX. Ventures in Transpersonal Healing: Living with Multiple Realities

1. Maslow, Abraham H. *Motivation and Personality.* New York: Harper & Row, 1954.
2. Tart, Charles. *States of Consciousness.* New York: E. P. Dutton, 1975.
3. Green, Elmer and Alyce. *Beyond Biofeedback.* New York: Delacorte Press, 1977.
4. Grof, Stanislav. *Realms of the Human Unconscious.* New York: Viking, 1975.
5. Wilbur, Ken. *The Atman Project: A Transpersonal View of Human Development.* Wheaton, IL: Theosophical Publishing House, 1980.
6. Tart, Charles. "Transpersonal Realities or Neurophysiological Illusions?" In *Metaphors of Consciousness*, edited by Ronald S. Valle and Rolf von Eckarsberg. New York: Plenum Press, 1981.
7. Grof, Stanislav. "Modern Consciousness Research and Human Survival." In *Human Survival and Consciousness Evolution*, edited by Stanislav Grof. Albany, New York: State University of New York Press, 1988.
8. Grof, op.cit.
9. Tart, Charles (1981), op. cit.

Chapter X. Fields in Which We Live: Toward a Comprehension of Energy Flow

1. Swami Rama. "Energy of Consciousness in the Human Personality." In *Metaphors of Consciousness*, edited by Ronald S. Valle and Rolf von Eckarsberg. New York: Plenum Press, 1981.
2. Kunz, Dora. *The Personal Aura.* Wheaton, IL: Quest Books, 1991.
3. Kunz, Dora. Personal communication, 1997.
4. Krieger, Dolores. *Accepting Your Power to Heal.* Santa Fe: Bear & Co., 1993, pp. 117–130.

Chapter XI. Energetic Forces at Work during the Therapeutic Touch Session

1. Puthoff, H. E. and R. Targ. "A perceptual channel for information transfers over kilometer distances: historical perspectives and recent research." *Proceedings of the Institute of Electrical and Electronic Engineers.* Vol. 64 (1976); 329–354.

2. Krieger, Dolores. "Visualization by Professional Nurses During Meditation on Hospitalized Patients Who Are at a Distance." Paper presented to Sigma Theta Tau, Alpha Ieta Chapter, University of Texas/Houston, April 10, 1981.

_______. *Living the Therapeutic Touch: Healing as a Lifestyle.* New York: Dodd, Mead & Company, 1987, pp. 81–102.

3. Tart, Charles. "Physiological Correlates of Psi Cognition." *International Journal of Parapsychology* 5 (1965): 375–386.

4. Yogi Ramacharaka. *Science of Breath.* Chicago: Yoga Publication Society, 1940, p. 17.

5. Kunz, Dora. *The Personal Aura.* Wheaton, IL: Quest Books, 1991.

6. Krieger, Dolores. *Living the Therapeutic Touch*, op. cit., p. 44.

7. Karagulla, Shafica, and Dora Kunz. *The Chakras and the Human Energy Field.* Wheaton, IL: Quest Books, 1989.

8. See "The Human Barrier Game" and "The Search for Ordering Principles" in *Accepting Your Power to Heal: The Personal Practice of Therapeutic Touch*, by Dolores Krieger. Santa Fe: Bear & Co., pp. 118–130.

9. Personal notes.

10. A good book to recommend to either the TT therapist or the healee is *The American Family Medical Guide*, edited by J. R. M. Kunz and A. J. Finkel (New York: Random House, 1987). Kunz and Finkel have done a very good job of using a logic tree to determine the nature of a large group of symptoms. Of particular note is the efficiency of their system for making the reader aware of emergency conditions.

11. Campbell, Joseph, and Bill Moyers. *The Power of Myth.* New York: Doubleday, 1988.

12. Swan, James A. *Sacred Places: How the Living Earth Seeks Our Friendship.* Santa Fe: Bear & Co., 1990.

13. Campbell, op. cit.
14. Swan, op. cit., p. 42.
15. Assagioli, Robert. *Psychosynthesis.* New York: Viking Press, 1971.
16. Kunz, Dora. *The Real World of Fairies.* Wheaton, IL: Theosophical Publishing House, 1999.
17. Needleman, Jacob. *Consciousness and Tradition.* New York: Crossroads, 1982.
18. See Credo Vusa'Mazulu Mutwa. *My People.* Trebridge, Kent: Peach Hall Works, 1969.
19. Sheldrake, Rupert. *A New Science of Life: The Hypothesis of Formative Causation.* Los Angeles: J. P. Tarcher, 1981.
20. Eliade, Mircea. *Patanjali and Yoga.* New York: Schocken, 1975, p. 38.
21. Targ, R., and H. E. Puthoff. "Information transfer under conditions of sensory shielding." *Nature* 252 (1974) 602–607.
22. Krieger, Dolores. "Visualization by Professional Nurses During Meditation on Hospitalized Patients Who Are at a Distance." Paper presented to Sigma Theta Tau, Alpha Ieta Chapter, University of Texas/Houston, April 10, 1981.
23. Boyd, Douglas. *Rolling Thunder.* New York: Delta, 1974.
24. Campbell, op. cit.

Chapter XII. The Warrior-Healer

Chapter XIII. To Be a Warrior-Teacher

Chapter XIV. Awareness of Learning Styles

1. Krieger, Dolores. *Therapeutic Touch: How to Use Your Hands to Help or to Heal.* New York: Simon & Schuster, 1979.
_______. *Accepting Your Power to Heal: The Personal Practice of Therapeutic Touch.* Santa Fe: Bear & Co., 1993.
2. Braud, William. "Parapsychology and Spirituality." *ReVision*, No.1, Summer 1995, p. 40.

Appendices

1. From Dolores Krieger, "Therapeutic touch during childbirth preparation by the Lamaze method. Its relation to marital satisfaction and state anxiety of the married couple." (Nsg. Research Emphasis Grant for Doctoral Programs, USPS, #NU-oo833-02. 1983)

2. Ferguson, Cecilia K. "Subjective Experience of Therapeutic Touch: Psychometric Examination of an Instrument." University of Texas/Austin, 1986.

3. This means that SETTS could be used to distinguish experienced TT therapists from those therapists who had little experience to the extent that, in a random group of one hundred TT therapists, ninety-nine would be correctly classified.

INDEX

In this index definitions are indicated with italicized page numbers. Photographs and figures are indicated by the use of *f* following the page number.